Inorganic Polymers

Inorganic Polymers

N. H. Ray

ICI Corporate Laboratory
The Heath, Runcorn, Cheshire, England

1978

ACADEMIC PRESS

London · New York · San Francisco
A Subsidiary of Harcourt Brace Jovanovich, Publishers

U.K. Edition published and distributed by
ACADEMIC PRESS INC. (LONDON) LTD.
24/28 Oval Road
London NW1

United States Edition published and distributed by
ACADEMIC PRESS INC.
111 Fifth Avenue
New York, New York 10003

Library of Congress Catalog Card Number: 77-93487
ISBN: 0-12-583550-7

Printed in Gt. Britain by
Page Bros (Norwich) Ltd, Norwich

Preface

During the twenty years or so that followed the end of World War II, there was a rapid growth of interest in the synthesis of inorganic polymers, which was stimulated by the hope that new materials would be discovered with enhanced thermal stability and resistance to oxidation. Much research was done, not only within industry, but also in universities and government laboratories. In the event, so little emerged in the way of materials with real commercial value, that many people became disillusioned with the idea of inorganic polymers. Disappointment with the poor return from their outlay led most of the people who controlled the finances to withdraw their support from work in this area, and in the ensuing climate of opinion it became very unpopular to embark on any new line of investigation. In spite of this, we are faced with the situation that in the long term, the world's only inexhaustible resources are inorganic; ultimately, even renewable organic materials like cellulose and vegetable protein will have to be regarded primarily as sources of food and clothing, and their conversion into plastics will have to be severely restricted.

The purpose of this book is twofold; firstly it is an attempt to change current opinion about inorganic polymers and to regenerate interest in the subject, particularly in universities, by drawing attention to the existence of a wide variety of inorganic polymers with potentially useful properties, most of which have been largely ignored by polymer scientists because of their cross-linked structures; and secondly to indicate some of the directions in which future research might be encouraged, so that through a better understanding of these systems we might be able to utilize them more effectively.

N. H. Ray
May 1978

I.C.I. Corporate Laboratory,
Runcorn Heath, Cheshire

Contents

1

Introduction

A New Approach to Inorganic Polymers

The term "inorganic polymer" has been used with a variety of meanings. Currell and Frazer [1] defined an inorganic polymer as a macromolecule which does not have a backbone of carbon atoms; Holliday [2] accepts that substances such as diamond, graphite, silica and inorganic glasses are inorganic polymers, pointing out that even concrete has been regarded as a polymer. Such a broad concept of what should be included in the class of inorganic polymers is by no means generally accepted; some scientists prefer to restrict the term "polymer" to essentially linear macromolecules. For example, MacCallum[3] chose to restrict a review of inorganic polymers to linear polymers having at least two different elements (one of which may be carbon) in the backbone of the repeat unit. The justification for this restriction was that in most cases such polymers are soluble and therefore more easily characterized. This is understandable, because polymer science has developed almost entirely from studies of organic polymers, most of which are linear macromolecules that are soluble and thermoplastic. In fact, for most people the word polymer means substances like polythene, nylon, "Perspex" and "Terylene". As a result of intensive commercial development, synthetic organic polymers have not only been able to satisfy many everyday needs but have also made it possible to obtain effects that could not have been obtained in any other way. Nevertheless, there still remain a number of practical requirements they cannot fulfil. These are principally concerned with thermal stability and oxidation resistance, including resistance to fire. The rapid growth of the aerospace industry following the 1939–1945 World War created an urgent demand for new, non-metallic materials that could retain their mechanical and electrical properties at higher temperatures than most organic plastics can withstand. This stimulated the interest of many chemists throughout the world to investigate the possibility of synthesizing inorganic polymers with the required physical and mechanical properties. As a result, much effort was expended during the succeeding 25 years or so in attempts to synthesize long-chain polymers from inorganic elements, generally following the structural patterns of the most useful organic polymers, in the misguided belief that this was the best way of achieving the required combination of thermoplasticity and thermal stability. In order to achieve this kind of structure much attention had to be directed to solving the problem of preventing the polyvalent inorganic elements such as boron, aluminium, silicon and phosphorus from reacting in a polyfunctional manner and producing branched and cross-linked polymers. This problem is aggravated by the fact that nearly every substituent that can reduce the functionality of inorganic monomers in polycondensation reactions (most inorganic polymers are, of necessity, polycondensation products) is either thermally unstable, or else unstable to hydrolysis or to oxidation. For example, to obtain a linear borate polymer with the repeat unit —O·B(R)—O—, the choice of the substituent R lies between hydrocarbon radicals such as alkyl or aryl, alkoxy and phenoxy radicals, and halogen or pseudo-halogen substituents. The dilemma arises from the fact that hydrocarbon substituents attached directly to boron are, with few exceptions, unstable to oxidation, and the exceptions are thermally unstable; on the other hand all the alternatives such as alkoxy, phenoxy, and halogen substituents are unstable to hydrolysis. The problem disappears, however, if the natural tendency of polyvalent inorganic elements to form highly cross-linked network structures is turned to advantage instead of being seen as a difficulty. This depends upon the possibility of developing suitable techniques for forming and processing cross-linked polymers, as well as demanding a much better understanding of the fundamental relationships between structure and properties in such materials.[4]

Recently, the incentive for this has greatly increased because during the last ten years some of the previously accepted defects of organic polymers have come to be regarded as more serious than was originally thought. The most important of these is the inherent flammability of organic plastics. Incidents such as the Summerland disaster in the Isle of Man on 2 August, 1973 have caused great concern about the safety of structures in which plastics are used extensively, and in many countries the ensuing legislation is already severely restricting the future use of plastics in public buildings and their furniture. This trend is certain to continue in spite of the development of more effective fire-retardant additives for plastics, which can only postpone the final issue, because no additive can alter the fact that once a fire has started, any organic polymer within range must in the last resort become a source of additional fuel. The use of many of the more effective fire-retardants can even be undesirable because plastics containing them are liable to evolve toxic fumes. A second drawback of organic plastics is their dependence on fossil carbon sources as raw materials. Although the eventual exhaustion of the earth's oil and gas deposits may still be two or three decades away, all of the industrial countries of the western world have already been affected by dramatic increases in the price of crude oil, and the consequent increases in the costs of most organic plastics. For these reasons, many people are starting to look for alternative materials that can be derived from the virtually unlimited mineral resources of the earth. While traditional materials like sand, cement, bricks and concrete will no doubt continue to be used for a long time to come, even if only because of the conservatism of local authorities, the rising cost of the labour needed to process and form them on site will gradually compel the adoption of other, more highly mechanized methods of construction, and there will be an increasing demand for new inorganic structural materials that can be pre-formed like plastics. Factors such as these are going to bring about a renewed interest in processing inorganic polymers, and such an approach to the subject must also take account of existing materials. It is for this reason that this book is strongly biassed towards naturally occurring inorganic polymers and synthetic materials of similar structure.

Definition of Inorganic Polymers

Anderson[5] said, with justification, that much of inorganic chemistry is the chemistry of high polymers; this is because many inorganic compounds exist only in the solid state as three-dimensional networks or two-dimensional layer structures. But if we were to classify as polymers all substances composed of giant molecules, including those that exist only in the solid state, our definition would have to include all the solid metals and ionic crystals, as well as many composite materials such as ceramics. Such a definition evidently embraces too wide a field to be useful, and it is obvious that certain restrictions must be imposed. What is the best kind of definition? From the scientist's viewpoint, the most satisfactory definition is one which, at least in a sufficient number of cases, can be tested experimentally and does not rely entirely upon some prior knowledge of structure. It would be reasonable to distinguish between substances that continue to exhibit the characteristic properties of macromolecules after some physical change such as melting or dissolving (even if the molecular weight or degree of cross-linking may have altered), and those that do not. For example, molten silica and the oxide glasses (silicates, borates and phosphates) have viscosities in the region of 10^5–10^6 N s m^{-2}, whereas the melt viscosity of nearly all metals and ionic salts are round about 10^{-3}–10^{-2} N s m^{-2}. All these substances consist of giant molecules in the solid state, but in this group only silica and the oxide glasses retain the characteristic properties of a macromolecular substance after melting and can therefore justifiably claim to be classified as polymers. Similarly, compounds such as sodium silicate and aluminium phosphate that dissolve in water and phosphoric acid respectively to give highly viscous solutions must be regarded as polymers, even though the molecules present in such solutions are of lower molecular weight and lesser complexity than those in the solid. By contrast, salts like zinc chloride and cadmium iodide, which form macromolecular crystals consisting of infinite two-dimensional sheets of atoms, are not polymers because when dissolved in water their solutions are not significantly more viscous than water. Evidently an experimental test of this sort cannot be applied to substances that neither melt nor dissolve without decomposition, that is without undergoing a chemical change apart from simple depolymerization. For such materials, however, a distinction can sometimes be made on the basis of vapour pressure. A true polymer does not exert an equilibrium vapour pressure, but only a dissociation pressure resulting from gradual degradation or depolymerization,[6] and this can be distinguished from a true vapour pressure by measurements of the rate of weight loss *in vacuo* at constant temperature. The rate at which a substance loses weight by simple volatilization does not vary with time, but the rate at which a polymer loses weight at a constant elevated temperature is time-dependent. This is because it is the result of a stepwise process, of which the successive steps that together make up the degradation or depolymerization reaction reach a state of dynamic equilibrium only after a finite amount of degradation has taken place, and the appropriate amounts of intermediate products have accumulated. This time may be comparatively short, or it may be so long that a constant rate is never attained. For example, Hsieh[7] has distinguished in this way between the polymeric and non-polymeric forms of arsenious oxide; arsenolite, a molecular crystal composed of As_4O_6 units, loses weight at a constant rate when heated in a vacuum at 95°C; whereas claudetite, which is a high polymer made up of layers of AsO_3 units sharing oxygen atoms, shows a time-dependent rate of weight loss under similar conditions. The conclusion that claudetite is polymeric is confirmed in this instance by its high melt viscosity (10^5 N s m^{-2} near the melting point).

The essential difference between compounds that have quasi-polymeric structures that exist only in the solid state, and true polymers that retain at least some of the characteristic properties of macromolecules both in solution and in the molten state, is that the latter consist of molecules built up of mainly covalent bonds. We shall therefore define an inorganic polymer as a substance with giant molecules composed of atoms other than carbon and linked together by mainly covalent bonds. With this definition we will clearly include the elements sulphur, selenium and tellurium, and the chalcogenide glasses; silica, phosphoric oxide and boric oxide and the oxide glasses; the majority of silicate minerals; and substances like alumina and beryllium fluoride which are highly viscous in the molten state, though not in solution. It will be reasonable to regard cement as a polymer because its setting is a polycondensation reaction (see Chapter 7), but in this book some substances that used to be considered the most important inorganic polymers, like the polyphosphate esters and the organopolysiloxanes, which have a high proportion of organic substituent groups, will not be discussed.

Special Characteristics of Inorganic Polymers

Compared with most organic polymers, inorganic polymers are generally stronger, harder, more brittle and usually insoluble. With few exceptions (for example sulphur) they do not burn and only soften or melt at very high temperatures. These characteristic features of high modulus, lack of ductility, insolubility and inertness to heat are typical of polymers with a highly cross-linked network, and many of the properties of inorganic polymers can be ascribed to their cross-linked structure.

Modulus of Elasticity

Mechanical properties such as Young's modulus vary considerably from one polymer to another, but the range of values that includes the elastic moduli of nearly all organic polymers lies below the range of moduli encompassed by most typical inorganic polymers. For instance, the Young's moduli of silica, the polymeric silicates and in fact the majority of all types of oxide glasses are greater than 40 GN m^{-2}, while with few exceptions the Young's moduli of organic polymers are less than 15 GN m^{-2} (Table I).

Holliday (loc. cit.) has drawn attention to the existence of a rough correlation between the Young's modulus of a polymer and the relative number of network bonds per unit volume. It has also been observed that the modulus of a wide variety of multicomponent oxide glasses derived from a common parent network is approximately proportional to the average number of oxygen atoms (and hence of network bonds) in unit volume.[8] The underlying basis of these relationships is not difficult to understand. A polymeric solid can be considered as an aggregate of macromolecular segments that are held together by a variety of forces, ranging in strength from van der Waals forces and hydrogen bonds between adjacent portions of different molecules, to strong covalent bonds between segments of the same molecule and at cross-linking points. The elastic modulus of such an assembly of forces can be roughly estimated in much the same way as that of a composite material, such as a fibre-reinforced plastic, by taking the weighted average of the elastic moduli of the separate components, the appropriate weighting factors being their respective volume fractions. To a first approximation the relatively weak intermolecular forces can be neglected in comparison with the intramolecular covalent bonds, so that the Young's modulus of such an aggregate should be approximately proportional to the number of covalent network bonds per unit volume, multiplied by their force constants, the constant of proportionality being determined by the orientations of the bonds relative to the direction of stress (e.g. $2/\pi$ for bonds that are randomly oriented in all directions in a plane). Thus in a family of polymers built up of similar bonds,

Table I
Mechanical properties of some inorganic and organic polymers

Material	Young's Modulus $GN\,m^{-2}$	Fracture Stress $MN\,m^{-2}$	Failure Strain %	Density $g\,cm^{-3}$
Inorganics				
(1) Amorphous Polymers:				
Fused Silica	70	350–1500	0·5–2	2·2
Borosilicate glasses	68	700–1200	1–2	2·5
Borophosphate glasses	40–50	350–1000	1–2	2·9–3·5
(2) Crystalline Polymers:				
Asbestos	140–175	700–3000	0.5–1	2·6–3·4
"Saffil" alumina fibre	100	1000	1	2·5
"Saffil" zirconia fibre	100	700	0·7	5·4
Organics				
Cotton	5–11	300–800	6–7	1·5
Silk	8–13	300–600	4–5	1·3
Nylon	6–11	380–420	4–6	1·1
Terylene	12–15	750–1000	6–7	1·4

such as, for example, the silicate glasses, the modulus will be proportional to the volume fraction of Si—O bonds; that is, to the volume fraction of oxygen atoms. It also follows that the majority of inorganic polymers, which have a cross-linked network structure with a high density of covalent bonds, will be stiffer and harder than organic polymers with a linear chain structure, unless the latter are specially oriented so that all the bonds lie along the direction of stress.

Tensile Strength and Brittleness

Brittleness is another characteristic property of polymers which have highly cross-linked networks. A material is described as brittle if it fractures at very low values of strain; for example when the fracture stress is less than 1 or 2% of the modulus. Ductile and plastic materials like polyethylene can extend by 20% or more before fracture, and tough materials like steel and cotton reach 5–10% strain at fracture; by contrast brittle materials such as glass and asbestos break at strains of <1% (Table I).

To relate these properties to structure, it is necessary to consider the ways in which plastic yield can occur in a polymeric solid. Strain is of two kinds: elastic strain which is completely reversible and returns towards zero as fast as the stress is removed; and plastic yield which is a combination of permanent deformation and a time-dependent strain that decreases slowly after the removal of stress. Clearly the structural changes that are concerned in elastic strain must be completely reversible, but plastic yield must involve a change of structure to a new, but at least temporarily stable, configuration. The attractive force between two atoms joined by a covalent bond at first increases as the separation is increased, reaching a maximum value when the bond length has extended by about 20%; then with further extension the interatomic force decreases fairly quickly. It follows that a force that is large enough to stretch a covalent bond by more than 20% is sufficient to cause rupture of the bond. Consequently, if a polymer with a cross-linked network structure has such a density of covalent bonds that no mechanical deformation can take place without stretching some of them, the maximum elastic strain at any point will be limited to 20%. In practice, of course, the maximum total strain will be much less than this because fracture will be initiated from any defect such as a surface crack where the local strain might reach 20%. Any greater strain must either produce permanent deformation involving bond relocation, or rupture. As the structure is made less complex and the density of covalent linkages is decreased, the segments of network that are unconstrained except by bond rotation become larger, and a point will be reached when some deformation can take place by rearrangement of bond angles alone without alteration of length. At this stage the polymer becomes significantly less brittle, at least for small strains. With only a few cross-links between polymer chains, considerable deformation can occur by the uncoiling of chain segments. For yield to be possible without fracture, the polymer structure must be such that it can change from one stable configuration to another, with a resultant increase in at least one dimension. In an assembly of linear macromolecules this can happen by the straightening of previously coiled chains, the relatively weak van der Waal's forces between adjacent chain segments being easily extended, severed and then relocated. But a three-dimensional network can only change to a new configuration by rupturing and reforming covalent bonds. Bond interchange can be induced by mechanical stress, but because of the magnitude of the mechanical equivalent of heat, the amount of energy that can be introduced in this way is strictly limited, and bond interchange can only be brought about by stress alone if the temperature is already high enough for the internal energy of the bonds involved to be nearly sufficient for bond interchange to occur spontaneously—that is near to the glass transition temperature. If the temperature is too low, the stress necessary to initiate bond interchange will be so large that it will exceed the fracture stress of the material. Thus at low temperatures—that is temperatures low in relation to the glass transition temperature—polymers with a three-dimensional network structure do not normally undergo plastic yield before fracture, and it is only at temperatures near the glass transition that a yield stress lower than the fracture stress can be observed. Since the glass transition temperature of a polymer with a cross-linked covalent network is necessarily much higher than that of a linear or mainly linear polymer, most inorganic polymers are brittle at ordinary temperatures. Because of the absence of plastic deformation at low temperatures, when an external stress is applied to a highly cross-linked network polymer, any minute cracks in the surface can act as stress-raisers, and the actual value of fracture stress is strongly dependent on the surface condition. This is because the externally applied stress is increased locally at the root of a crack by a factor that is inversely proportional to the radius of the tip of the crack. The sharper the crack, the greater the local stress, and when it exceeds the intrinsic strength of the material, the crack propagates uncontrollably, resulting in fracture.[9] In materials that are capable of yielding plastically under the prevailing conditions of temperature and stress, plastic deformation of the material in the immediate neighbourhood of the crack leads to an increase in tip radius and a consequent reduction in local stress; but in highly cross-linked network polymers this cannot occur, and fracture is observed long before the applied stress has reached the intrinsic strength of the network—usually at values as low as 0·5–1·0% of the Young's modulus.

Even with organic polymers that can yield a little before fracture—for example polymethylmethacrylate—the process of stress relaxation by plastic deformation is not instantaneous and the relaxation time is a function of temperature. Consequently at low temperatures and with rapid application of stress, brittle failure and notch sensitivity is observed. However, most organic polymers are comparatively tough and lacking in notch sensitivity at ordinary temperatures, because both their glass transition temperatures and yield stresses are comparatively low; by contrast, many inorganic polymers are notch-sensitive and fail by crack-propagation at stress levels that may be as much as two orders of magnitude lower than the intrinsic strength of the polymer network. By careful preparation it is sometimes possible to obtain specimens that are nearly free from surface defects; specially drawn fibres usually offer the most promise in this direction, although chemical polishing by reagents that selectively dissolve sharp corners and acute angles–hydrofluoric acid treatment of silicate glasses is a good example–can provide bulk polymers with surfaces that are equally free from imperfections. Under these conditions, values of fracture stress approaching 10% of the elastic modulus have been observed, and results like these show that the intrinsic strength of some inorganic polymers is as much as an order of magnitude greater than that of all but a few organic polymers, even though the potential strength is rarely, if ever, realized in practical situations.

Solubility

The dissolution of a solid in a solvent is a process that involves the insertion of solvent molecules between adjacent molecules of the solid, so that the intermolecular forces are reduced in strength sufficiently for whole molecules, molecular fragments, or aggregates of molecules to become detached from the remaining solid and dispersed in the solvent. Complete detachment of molecular fragments from a cross-linked polymer is impossible without structural degradation because, apart from gross defects such as voids and cracks, any intact piece of such a polymer must be a single giant molecule in which every part is connected through a continuous network of covalent bonds to every other part. When the cross-link density is low enough for there to be a significant proportion of flexible chains between the more rigidly bonded network points, solvent molecules can enter between chain segments and cause the polymer to swell and lose rigidity, but it cannot dissolve in the solvent. Whether the polymer actually dissolves or merely swells by imbibing solvent, if it is a true solvent, its subsequent removal—for example by evaporation—must restore the polymer to its original state. This is the typical behaviour of both linear and cross-linked organic polymers in which the density of cross-linking is not too high, and which are chemically inert to the solvent in question. The situation is different with the majority of inorganic polymers for two reasons: firstly the chain segments between cross-linking points are short and comparatively stiff (in silica, for example, there are cross-links at every alternate atom along a chain) and consequently there is not enough flexibility in the structure to admit intercalated solvent molecules; secondly, most inorganic polymers are built up of highly polar repeat units, and the only liquids that are sufficiently polar to be capable of dissolving them often react chemically with them. For these reasons few inorganic polymers either dissolve in the true sense, or swell reversibly by imbibing solvent molecules. Some apparent exceptions to this generalization prove to be otherwise on closer examination. The simple alkali silicates appear to dissolve in water without change, but there is ample evidence to show that the molecules present in such solutions are different from and much smaller than the molecules of which the solid is composed or those which are present in the molten silicate. Another apparent exception is the group of highly swollen aquagels formed by the high molecular weight alkaline earth polyphosphates such as magnesium metaphosphate; when these gels are allowed to dry out completely their structure alters and they will then no longer swell in water, showing that the swollen polymer must be different from the anhydrous material.

Glass Transition Temperature

The behaviour of polymeric materials on heating determines whether and how they can be processed into useful shapes, what is their useful temperature range, and at what temperature depolymerization or thermal degradation sets in. Most polymers have a second-order transition at a temperature that is known to polymer scientists as the glass transition temperature, and to glass technologists as the transformation temperature; a knowledge of the glass transition temperature is a first step towards a quantitative description of the thermal behaviour of any polymeric material. The glass transition temperature can be defined as the temperature (or temperature range) above which the stress–strain relationship in the material becomes time-dependent, and all stresses will eventually relax to zero within a finite time; below the glass transition temperature the elastic part of the strain produced by any given stress is time-independent for reasonable experimental times. When a polymer is heated through the glass transition range, the internal energy of the molecules and the resulting thermal motion of the chain segments increase to the point at which movement of chains past one another can take place under infinitesimally small stress. The temperature at which this will occur depends on the strength and number of interchain linkages, and will be much lower for linear-chain hydrocarbon polymers, with only van der Waals forces between chain segments, than it is for cross-linked materials. The glass transition temperature will be higher for polymers that have covalent cross-links than it is for those with hydrogen bonds between chains like polyamides and polyesters, and will increase with the density of cross-linking. The effect of cross-link density on the glass transition temperature has been studied by di Marzio and Gibbs[10] who showed that the glass transition temperature of a cross-linked polymer is related to the cross-link density (defined as the fraction of repeat units that are linked to more than two others) by the equation

$$T_x/T_0 = 1/(1-Kx),$$

where T_0 is the glass transition temperature (in K) of the corresponding linear polymer, T_x the glass transition temperature of the cross-linked polymer with cross-link density x, and K is a constant that to a first approximation is independent of the material. This relationship was confirmed by experimental measurements for organic polymers with comparatively low values of cross-link density; and it predicts that the glass transition temperature will rise very steeply once the cross-link density increases above the level at which $2Kx - K^2x^2$ is small in comparison with unity, for

$$\mathrm{d}T_x/\mathrm{d}x = KT_0/(1-Kx)^2.$$

With organic polymers this means that if there is more than about 5% of cross-linking the glass transition temperature is very likely to exceed the decomposition temperature of the polymer, and consequently the polymer degrades or decomposes before it begins to soften. This is why cross-linked organic polymers are unprocessable. The situation is quite different with inorganic polymers, many of which easily withstand heating to temperatures well in excess of their glass transition; oxide glasses, for example, with highly cross-linked network structures, are commonly processed at temperatures that may be 200–400°C higher than their glass transition temperature. It has been shown that di Marzio's equation relating glass transition temperature to cross-link density also holds for polyphosphate glasses up to cross-link densities of at least 0·5, that is with one cross-link for every two repeat units.[11] At such high cross-link densities the mechanism of flow (stress relaxation) must be different from that operating in carbon polymers; the most likely mechanism is a process involving bond interchange (Fig. 1.1). As Eisenberg[12] has observed, bond interchange is

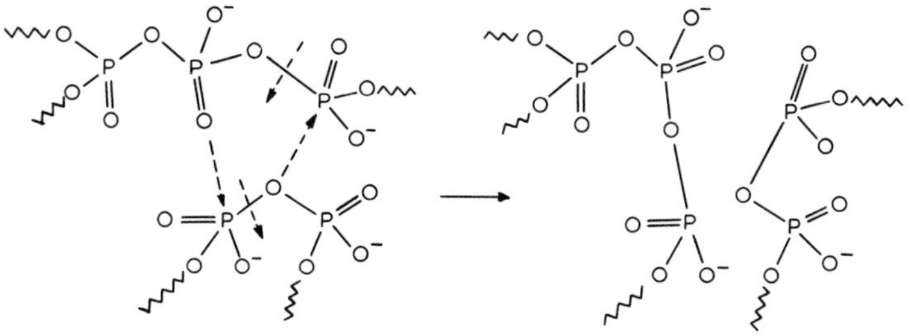

Fig. 1.1. Bond interchange in a cross-linked polyphosphate.

common in inorganic polymers in the molten state, and in certain cases (notably sulphur and linear polysilicates) it even appears to provide the mechanism of stress relaxation in linear polymers. The ability of inorganic polymers to undergo bond interchange is one of the most fundamental differences between organic and inorganic polymers. Once this phenomenon is properly understood, it will be evident that the problem that engaged the attention of so many synthetic polymer chemists for so long, that of limiting the functionality of the inorganic polymer-forming elements to two so as to produce only linear chains, disappears entirely. By taking advantage of the phenomenon of bond interchange it becomes possible to design highly cross-linked inorganic polymers with relatively low glass transition temperatures, and with sufficiently flexible chain segments between cross-linking points to accommodate a certain amount of plastic yield and thereby reduce brittleness. By way of illustration, there are a number of cross-linked polyphosphates that have glass transition temperatures as low as those of some typical organic thermoplastics[13] and there are some lightly cross-linked arsenic chalcogenides that exhibit plastic yield.[14]

Crystallinity

While few organic polymers are 100% crystalline, there are many which contain both crystalline and amorphous regions within the same, chemically homogeneous material. The relative proportions of crystalline and amorphous phases vary with the rate of cooling of the polymer and are reflected in the density and the mechanical properties of the material.

Among inorganic polymers, on the other hand, 100% crystallinity is quite common, and several well-known inorganic polymers occur in more than one fully crystalline form: quartz, cristobalite and tridymite are all different crystalline varieties of the same polymer, which can also be obtained in the wholly amorphous state as vitreous silica; again, Kurrol's and Maddrell's salt are different crystalline forms of the same linear high molecular weight polyphosphate.[15] Many inorganic polymers, such as the chalcogenide glasses, are wholly amorphous, but partially crystalline inorganic polymers are comparatively rare except among chemically heterogeneous materials like ceramics and cement, in which the crystalline and amorphous phases have different chemical compositions. Even in a partially crystalline glass like "Pyroceram", a typical glass-ceramic, the crystalline and glassy phases are of different composition.

Methods of Characterization

The traditional methods of polymer physics that have been used for characterizing organic polymers depend to a large extent upon the possibility of making measurements of their properties in solution: for example the relationships between viscosity and concentration, and the scattering of light by polymer solutions can give information about the size and shape of the dissolved polymer molecules; the partition of a soluble polymer between a solute and a porous gel, or between two different solvents, provides information about the distribution of molecular sizes. These techniques cannot generally be applied to inorganic polymers, because the majority are insoluble. Much less attention has been given to the characterization of insoluble polymers, the structure of which can only be deduced from measurements of their properties in the solid state, supplemented by measurements such as viscosity of the melt where thermal stability and melting temperature allow. So far as cross-linked polymers are concerned, classical polymer science has made very little contribution to the understanding of the relationship between structure and properties: but the polymer technologist concerned with processing has to study melt properties and the engineer who wishes to utilize polymeric materials has to study solid state properties. Both have made significant contributions to the development of techniques for the characterization of insoluble and cross-linked polymers. Many inorganic polymers that are cross-linked and insoluble, nevertheless melt without decomposition, so that their behaviour in the glass transition range and in the molten state can be a valuable source of information about their structure. Differential thermal analysis, high-temperature viscosity measurements, and Raman and infra-red spectroscopy in both the solid and molten states can yield essential information about the types of bonding and relaxation mechanisms present in inorganic polymers. Recently, the application of nuclear magnetic resonance measurements in the solid state, and of X-ray scattering to amorphous materials, has yielded important structural information about inorganic glasses. Although many problems remain to be solved, it can be said that almost as much is now known about the structure of some inorganic network polymers such as silica as about the majority of linear organic polymers.

Classification of Inorganic Polymers

The purpose of any classification should be firstly to economize on descriptive text by bringing together those items with similar properties or behaviour, for most of which a common description can serve; and secondly to enable useful predictions to be made about new entries that might subsequently have to be included in the scheme. To succeed in the latter aim, it must be possible to classify any new item without having too much prior knowledge of its properties or behaviour. The periodic classification of the elements is a good example of this, because it makes it possible to describe many of the properties of, for example lithium, sodium, potassium, rubidium and caesium in a single set of statements about the alkali metals. Furthermore, when a new element with a similar type of electronic structure was discovered, the main essentials of its chemistry could be predicted with some confidence.

Inorganic polymers, unlike organic polymers, can contain a wide variety of elements, and in the past it has been customary to classify them on the basis of composition. Such a scheme has some merit in discussing chemical properties and methods of synthesis, but from the standpoint of polymer science, a classification based on structural type has the advantage of bringing together materials with similar physical properties, particularly those properties which affect their utility and methods of processing. Since many inorganic polymers have a network structure, in this book it is proposed to adopt a classification based on the idea of network connectivity. For each class of polymer there is a real or notional parent network from which the family of real polymers can be derived; for example, all the silicate glasses can be considered to be derivatives of the parent network present in amorphous silica. The connectivity of the parent network can be defined as the number of network bonds that link each repeat unit in the network: thus silica with the tetravalent repeat unit SiO_4 has a connectivity of 4, while boric oxide with the trivalent repeat unit BO_3 has a connectivity of 3, and a linear polymer such as polymeric sulphur has a connectivity of 2 (Fig. 1.2). Polymers that are derived from the parent networks by partial substitution

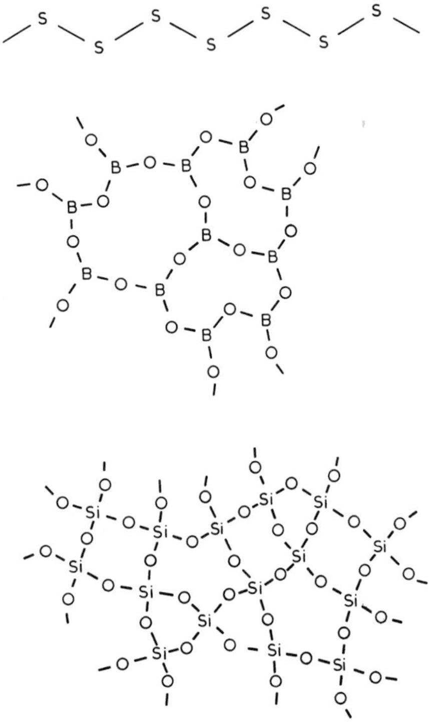

Fig. 1.2. Inorganic polymers with connectivities of two (sulphur), three (boric oxide), and four (silica).

with metal cations, hydroxyl groups, or organic ligands will have an average connectivity lower than the parent, and it is convenient to distinguish the actual frequency of linkages in any particular derived polymer from the connectivity of the parent network (which represents the maximum attainable for that class of polymer) by reference to the cross-link density, which has been defined by Flory [16] as the fraction of monomer (i.e. repeat) units that are cross-linked. For an adequate description of very highly cross-linked systems, such as the silicates, in which there can be up to two cross-links per repeat unit, Flory's definition requires some modification because strictly speaking the "fraction of repeat units" cannot be greater than unity. However, since the connectivity of the parent network is the maximum attainable cross-link density, it is evident that the cross-link density of any derived structure with lower connectivity than the maximum can logically be defined as the fraction of the parent connectivity that has been achieved. This is the same as the actual number of cross-linkages in a section of network divided by the maximum possible number for that network, averaged over a large number of repeat units. Within a given class of network polymers we shall regard the connectivity as a constant which characterizes the parent, fully cross-linked, network, and the cross-link density as a variable that describes how closely the structure of any member of the class approaches that of the parent network. Thus a highly condensed polyphosphoric acid containing, say, one hydroxyl group for every two phosphorus atoms, can be classified as a derivative of phosphoric oxide, that is as a three-connective polymer with a cross-link density of 0·5. Similarly, an alkali silicate glass with the molar ratio of silica to soda, $SiO_2:Na_2O = 3$ would be classed as a derivative of silica, making it a four-connective polymer with a cross-link density of 0·67 (Fig. 1.3).

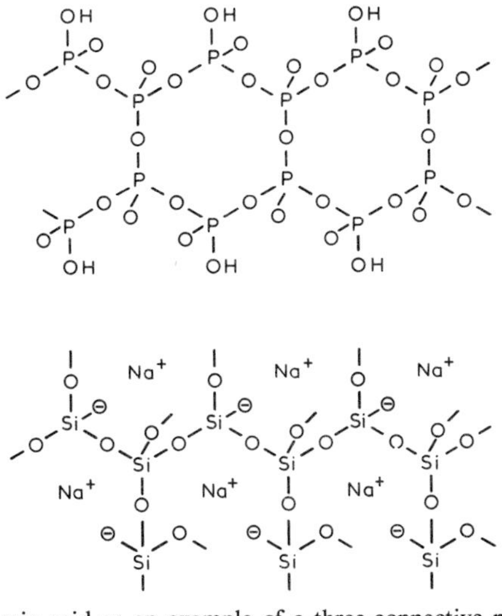

Fig. 1.3. Polyphosphoric acid as an example of a three-connective polymer with a cross-link density of 0·5, and sodium silicate as a four-connective polymer with a cross-link density of 0·667.

The main advantage of this scheme of classification is that polymers with similar physical and mechanical characteristics and areas of use fall into the same class. For example, all the two-connective (i.e. linear) polymers are soluble; with few exceptions they have low softening points, and many of them are flexible. The three-connective polymers are insoluble, rigid solids or glasses of intermediate softening point, while the four-connective polymers are generally brittle, high-modulus glassy or crystalline materials of very high softening point.

Networks of mixed connectivity exist, and the same principles can be applied. For example, silicon phosphate (SiP_2O_7) has a network structure consisting of two kinds of repeat units with connectivities six and four respectively, in which each silicon atom is linked to six oxygen atoms and each phosphorus to four (Fig. 1.4). The "average" connectivity of this network—if the concept had any meaning—would be 4·667, which is indicative of its extremely high melting point (>1200°C) and great chemical stability.

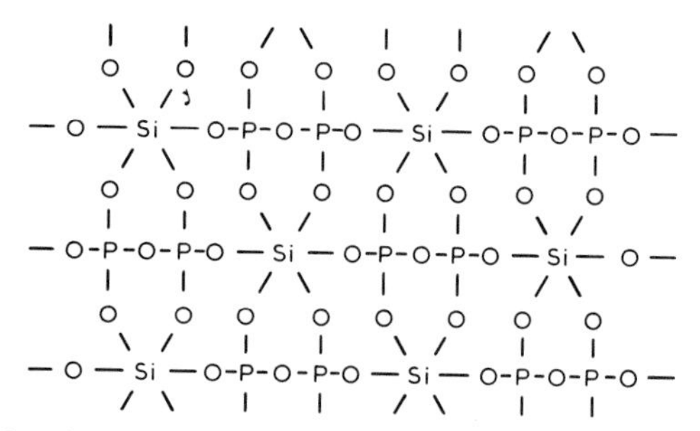

Fig. 1.4. Silicon phosphate as an example of a network with mixed connectivity (four and six).

Each network class contains several families of polymers, which may be further classified into crystalline polymers and amorphous polymers, as indicated in Fig. 1.5. Finally, particular polymers are conveniently grouped together on the basis of chemical composition; but from the viewpoint of polymer science, the conventional classification by chemical type may sometimes be a disadvantage, because it can conceal relationships which are common to that class of network and almost independent of its chemical nature. This is especially true of one important group of amorphous network polymers, the oxide glasses. These are hetero-atom networks composed of oxygen atoms linked together by mainly covalent bonds to atoms of multivalent non-metallic elements, such as boron, silicon, germanium, phosphorus and arsenic. They can also contain metal cations which are distributed throughout the network and are associated with singly-bonded oxygen atoms carrying a negative charge, so that the cross-link density of the polymer is reduced from that of the parent network in proportion to the mole fraction of cations. The customary classification of oxide glasses into silicates, borates, phosphates, etc., draws the attention away from the one feature that is common to them all, namely that these are all

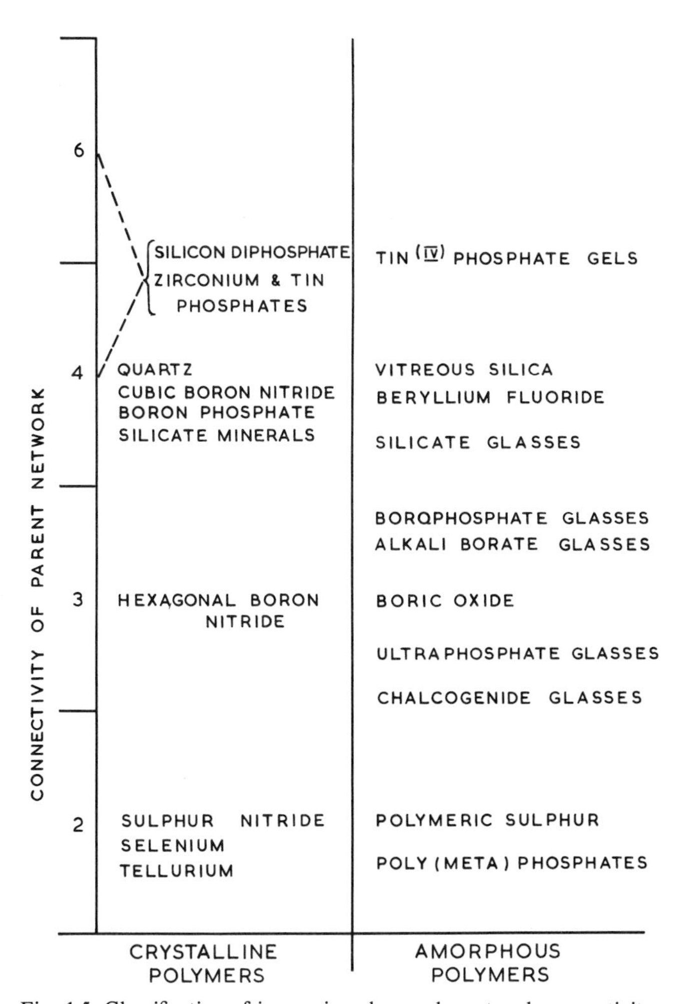

Fig. 1.5. Classification of inorganic polymers by network connectivity.

oxygen networks. The packing of an oxygen network is not greatly affected by the nature of the linking atoms, because they generally occupy a much smaller fraction of the total volume than the oxygen atoms. The weight of oxygen in unit volume of many oxide glasses does not vary greatly in spite of large differences in actual density and network connectivity, and the average value is remarkably close to the actual density of liquid oxygen at $-183\,°C$ (Table II). This common feature of oxide glasses is manifested in a number of physical properties, for example the Young's modulus, and the melt viscosity at the melting temperature of the corresponding crystalline compound. The Young's modulus of oxide

Table II
Density, oxygen density and glass transition temperatures of polymeric oxide glasses

Network Connectivity	Oxide glass	Density $g\,cm^{-3}$	Oxygen density $g\,atom\ litre^{-1}$
3	As_2O_3	3·701	56·2
	B_2O_3	1·812	78·0
	P_2O_5	2·235	78·7
4	GeO_2	3·628	69·5
	SiO_2	2·203	73·5
	liquid oxygen (at −183°C)	1·149	71·8

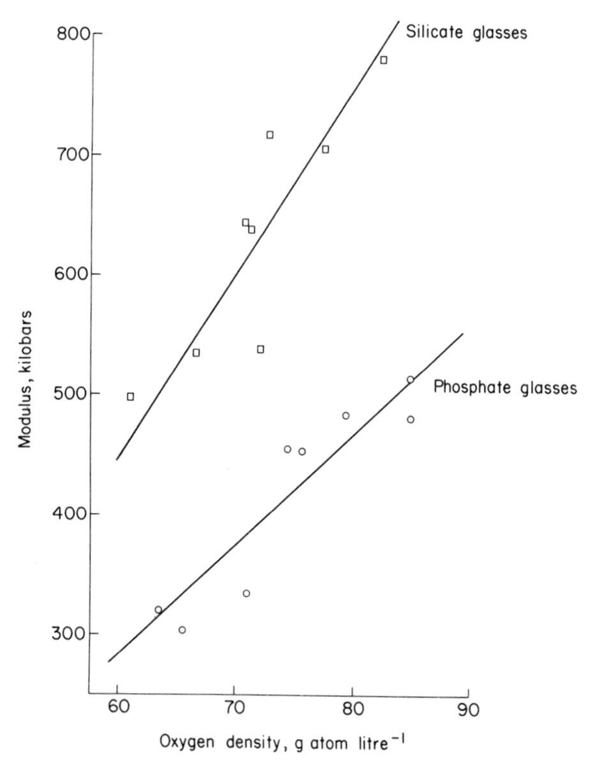

Fig. 1.6. Young's modulus of two series of oxide glasses of different connectivity, plotted against oxygen density. (From Ray, N. H. (1974). *J. Non-cryst. Solids,* **15**, 432. With permission.)

glasses of similar network connectivity is approximately proportional to their oxygen density (weight of oxygen in unit volume), and at similar oxygen densities the modulus is roughly proportional to the connectivity of the network (Fig. 1.6). Similarly, the melt viscosities of simple oxide network polymers all lie in the range 10^4–10^6 N s m^{-2} at the crystalline melting point, and the changes in melt viscosity that result from the introduction of metal cations into the network parallel the consequential changes in oxygen density.

History of Inorganic Polymers

The oldest synthetic inorganic polymer—for that matter the oldest of all synthetic polymers—was undoubtedly a simple alkali silicate glass used during the Badarian period in Egypt (*c.* 12 000 B.C.) as a glaze that was applied to steatite after it had been carved into the shapes of amulets, beads, and scarabs. According to Petrie,[17] the discovery of glazing was most probably accidental. A later development of this process resulted in the manufacture of faience, which is a composite material consisting of a core of powdered quartz or steatite covered with a comparatively thick layer of fused, opaque glass. Faience was used extensively from about 9000 B.C. onwards[18] to make beads, small figurines, and decorative objects. Glass proper probably first appeared around 5000 B.C., and the earliest known piece of glass that has been reliably dated is the lion's amulet originally found at Thebes, and now to be seen in the British Museum. This is an opaque blue glass with a partial coating of a transparent dark green glass that was made in 3064 B.C. Transparent, homogeneous silicate glasses first appeared around 1500 B.C.[19] In Egypt, glass was first used for vessels during the reign of Tuthmosis III (1900 B.C.), while the artistic use of glass for personal ornamentation in ear-rings, necklaces, pectorals and other adornments reached a state of perfection during the time of Tutankhamen (1300 B.C.). Several fine glass ornaments and jewellery were found in Tutankhamen's tomb, including two very fine bird's heads made of light blue glass and incorporated into a gold pectoral that was worn by the Pharaoh.

The manufacture of glass proper probably began in Syria, and a clay tablet dating from 1700 B.C. was found at Tel-Umar bearing an inscription which gives a recipe for the preparation of glass. Another clay tablet found in Mesopotamia and dated not later than 1600 B.C. carried a detailed recipe for the production of glaze. The technology of glass-making spread from Syria to Egypt and Palestine, and eventually throughout the Western World.

The next major discovery in the field of synthetic inorganic polymers occurred about 3000 years later in 1833, when Thomas Graham published the first accurate description of both crystalline and amorphous sodium polyphosphates.[20] The crystalline variety discovered by Graham is now known by the name Maddrell's salt; another crystalline form, known as Kurrol's salt, was first described by Tamman in 1892.[21] In 1891, Acheson discovered how to make silicon carbide[22] and started an industry that is still thriving. The oldest synthetic inorganic elastomer, polydichlorophosphazene, was first prepared by Stokes in 1897[23] and its rubber-like properties, unusual in an inorganic substance, have been the stimulus for many subsequent investigations of phosphazene polymers; up to the present time, however, no commercially successful product has yet emerged from the research in this field. By contrast, the monumental work on organosilicon compounds, for which F. S. Kipping will long be remembered, laid the foundations of a successful industry based on the manufacture of polysiloxane fluids, resins and elastomers. The prototype of these organo-inorganic polymers was first prepared by Kipping in 1904,[24] but it was not until after the 1939–45 World War that the polysiloxanes became available commercially. During the ensuing thirty years, scientific interest in the possibilities of synthetic inorganic polymers at first grew rapidly in the hope of fulfilling the needs of the space age for new, heat-resistant materials. Perhaps the most important discovery during this period—though not the outcome of polymer science, but rather that of glass technology—was that of glass-ceramics in 1947.[25] These are opaque, two-phase glasses of high mechanical strength and exceptional resistance to thermal shock produced by devitrification of a glass in which microphase separation has occurred. During the late sixties and early seventies there was a decline in the amount of effort directed towards the synthesis of long-chain inorganic polymers, probably because so much of this work has been unfruitful, and a relatively smaller number of investigators began to turn their attention to the utilization of naturally occurring inorganic polymers such as the silicate minerals. Possibly the most important advance in the field of synthetic inorganic polymer chemistry during this period was the discovery by Labes in 1973[26] that polymeric sulphur nitride (first reported in 1910 by Burt[27]) is a metallic-type conductor, which was found by Greene and others[28] in 1975 to become superconducting at sufficiently low temperatures. This is the first known example of a polymeric superconductor.

What of the future? It is the author's belief that the next major advance in inorganic materials is most likely to result from a careful study of the structure-property relationships among highly cross-linked polymers, such as the naturally occurring silicate minerals, and ultimately to depend upon the development of radically different methods of processing them.

References

1. Currell, B. R. and Frazer, M. J. (1969). *Roy. Inst. Chem. Rev.* **2**, 13.
2. Holliday, L. (1970). *Inorg. Macromol. Rev.* **1**, 3.

3. MacCallum, J. R. (1972). *In* "Kinetics and Mechanisms of Polymerization" (Ed. Solomon, D. H.) Vol. 3, Ch. 7, pp. 333–369. Marcel Dekker, New York.
4. Ray, N. H. (1975). *Endeavour* **34**, 9.
5. Anderson, J. S. (1961), "Introductory Lecture given at an International Symposium on Inorganic Polymers, Nottingham 1961". Chemical Society, London.
6. Livingston, H. K. (1970). *Inorg. Macromol. Rev.* **1**, 127.
7. Hsieh, J. T-T. (1949). MS Thesis. Wayne State University, Detroit.
8. Holliday, L. (1968). *J. Appl. Polymer Sci.* **12**, 333.
9. Hillig, W. B. (1962). *In* "Modern Aspects of the Vitreous State" (Ed. Mackenzie. J. D.) Vol. 2, p. 152. Butterworths, London.
10. di Marzio, E. A. and Gibbs, J. H. (1959). *J. Polymer Sci.* **40**, 121.
11. Ray, N. H. and Lewis, C. J. (1972). *J. Materials Sci.* **7**, 47.
12. Eisenberg, A. (1976). *Inorg. Macromol. Rev.* **1**, 75.
13. Ray, N. H., Lewis, C. J., Laycock, J. N. C. and Robinson, W. D. (1973). *Glass Technology,* **14**, 50.
14. Kurkijan, C. R., Krause, J. T. and Sigety, E. A. (1971). *IXth International Congress on Glass,* **1**, 4.
15. Elliott, J. H. (1975). "Ionic Polymers", (Ed. Holliday, L.). Ch. 7, p. 332. Applied Science Publishers, London.
16. Flory, P. J. (1953). "Principles of Polymer Chemistry" Cornell University Press, New York.
17. Petrie, W. F. (1930). "Arts and Crafts of Ancient Egypt". Foulis, London.
18. Morey, G. W. (1954). "The Properties of Glass", p. 4. Reinhold, New York.
19. Petrie, W. F. (1926). *J. Soc. Glass Technol.* **10**, 229.
20. Graham, T. (1833). *Phil. Trans.* **123**, 253.
21. Tamman, G. (1892). *J. prakt. Chem.* **45**, 417.
22. Acheson, E. G. (1893). *Chemical News,* **68**, 179.
23. Stokes, H. N. (1897). *Amer. Chem. J.* **19**, 782.
24. Kipping, F. S. (1904). *Proc. Chem. Soc.* **20**, 15.
25. Stookey, S. D. (1947). British patent 635 649.
26. Walatka Jr., V. V., Labes, M. M., and Perlstein, J. H. (1973). *Phys. Rev. Letters,* **31**, 1139.
27. Burt, F. P. (1910), *J. Chem. Soc.* 1171.
28. Greene, R. L., Street, G. B. and Suter, L. J. (1975). *Phys. Rev. Letters,* **34**, 577.

2

Linear (Two-connective) Polymers

POLYMERIC SULPHUR, SELENIUM AND TELLURIUM

The elements sulphur, selenium and tellurium are called the chalcogens because of the pronounced tendency of their atoms to link together to form long-chain homopolymers. The ability of sulphur to form homopolymers is second only to that of carbon, and gives rise to a wide variety of polysulphides and organic sulphur compounds containing chains of sulphur atoms. In spite of this tendency, high molecular weight straight-chain sulphur polymers are comparatively unstable, being readily depolymerized by heating, and there is no sulphur polymer that has achieved anything approaching the technical and commercial importance of the hydrocarbon polymers such as polyethylene. The only high molecular weight sulphur polymers that have found practical application on a sufficient scale to merit industrial production are the alkylene polysulphides, known as the "Thiokols", which have the general composition

$$[(CH_2)_x\text{—}S_y\text{—}]_n.$$

Since these substances contain a carbon chain in the backbone and are essentially organic polymers so far as their chemistry and physical properties are concerned, they will not be discussed here.

Polymeric Sulphur

The stable form of sulphur at ordinary temperatures is cyclo-octasulphur which melts at about 114°C (the exact melting point of sulphur always varies with the rate of heating, because of the comparatively slow transition from rhombic to monoclinic sulphur) to a liquid of quite low viscosity (about $1.1 \times 10^{-2}\,\mathrm{N\,s\,m^{-2}}$), indicative of a low molecular weight. As the temperature of the melt is raised, the viscosity at first decreases to about $7 \times 10^{-3}\,\mathrm{N\,s\,m^{-2}}$ at 150°C, and then between 159° and 165°C it increases very rapidly, reaching a maximum value of about $100\,\mathrm{N\,s\,m^{-2}}$ at 180°C. This change is due to a reversible ring-opening polymerization that involves the homolytic fission of the eight-membered ring to form a biradical, which can then either add to itself or react with another molecule of cyclo-octasulphur, so that the chain length increases by segments of eight atoms each time. As the temperature rises, the degree of polymerization increases rapidly and the weight fraction of polymeric sulphur also rises, so that the viscosity increases very rapidly from a low value corresponding to a monomeric liquid to the level typical of a molten polymer. As the temperature is raised further still, a ceiling temperature for chain growth is reached and although the fraction of polymer present in the melt continues to increase, the average molecular weight falls, so that the viscosity passes through a maximum value and then begins to decrease gradually. These changes have been studied by Gee,(1) Fairbrother, (2) and others. Tobolsky and Eisenberg(3) showed that the viscosity changes in both sulphur and selenium over the whole liquid range can be quantitatively explained in terms of the following sequence of reactions, where M stands for S_8:

$$M \overset{k_1}{\rightleftharpoons} \cdot M \cdot$$

$$M + \cdot M \cdot \overset{k_2}{\rightleftharpoons} \cdot M_2 \cdot$$

$$M + \cdot M_{n-1} \cdot \overset{k_3}{\rightleftharpoons} \cdot M_n \cdot$$

For both sulphur and selenium all the propagation steps are identical, except for chain length, and $k_n = k_2$. The number average degree of polymerization is given by

$$P = 1/(1 - k_2[M]),$$

so that for values of P that are large in comparison with unity, the concentration of monomer is given by $[M] = 1/k_2$ approximately. The amount of monomer present can be measured by extraction of the polymer with carbon disulphide, in which the polymer is insoluble. If $[M_0]$ is the total concentration of sulphur or selenium (i.e. 3.9 mol $\mathrm{Kg^{-1}}$ for S and 1.58 mol $\mathrm{Kg^{-1}}$ for Se)

$$[M_0] = [M] + k_1[M]/(1 - k_2[M])^2 = [M] + k_1[M]P^2$$

so that for values of P much larger than unity,

$$k_2[M_0] = 1 + k_1 P^2.$$

In this way, values of the equilibrium constants k_1 and k_2 can be determined by measuring the equilibrium polymer concentration and the average molecular weight at various temperatures. From these results, the entropy and enthalpy of the reactions can be deduced, and Eisenberg's values are given in Table III. One consequence of this model is that molten sulphur at any temperature should consist only of molecules containing multiples of eight atoms. Recently, however, Schmidt[4] has demonstrated the presence of small amounts of cyclododecasulphur, S_{12}, in the melt, showing that more complex reactions must be taking

Table III.
Entropy and enthalpy of polymerization of sulphur and selenium

Reaction	Sulphur		Selenium	
	H° kJ mol^{-1}	S° J mol^{-1}K^{-1}	H° kJ mol^{-1}	S° J mol^{-1}K^{-1}
Initiation	137·3	96·3	104·6	96·3
Propagation	13·3	19·4	9·5	22·9

place. Van Wazer[5] suggested that there may be "scrambling" reactions in the melt, just as there are in phosphate melts, for example, which could result in molecules of any chain length; in support of this Eisenberg has pointed out that the actual viscosity of molten sulphur is at least an order of magnitude less than would be expected from its molecular weight, indicating relatively easy bond interchange.[6] Nevertheless, the distribution of molecular sizes is not what might be expected from the statistics of a simple dynamic equilibrium between all possible species, because the amounts of cyclododecasulphur found in the melt are much too small for that.

Polymeric sulphur of high molecular weight can be isolated from the melt by quenching from 180°C into ice-water. When freshly prepared in this way, polymeric sulphur is an elastomer with a glass transition temperature of −30°C. Tobolsky[7] has shown that sulphur in this condition is elastomeric because it is plasticized by cyclo-octasulphur in a metastable liquid state; pure polymeric sulphur, obtained by thorough extraction of the plasticized material with carbon disulphide, has a glass transition temperature of 75°C. Eisenberg[8] studied the viscoelastic properties of plastic sulphur and found that a time–temperature superposition of stress-relaxation curves fails at long relaxation times; he concluded that stress relaxation in polymeric sulphur above the glass transition temperature must involve bond interchange. Eisenberg suggested that processes such as

$$\sim S\cdot + \begin{matrix} S\sim \\ | \\ S\sim \end{matrix} \rightarrow \sim S \begin{matrix} S\sim \\ / \\ \\ \cdot S\sim \end{matrix}$$

could occur, involving as a necessary initiation step the homolytic dissociation of an S–S bond into two free radicals. The activation energy for such a process can be estimated, by analogy with other polysulphides, to be of the order of 150 $kJ\,mol^{-1}$, and the activation energy for viscous flow in sulphur is actually 142 $kJ\,mol^{-1}$. This also provides an explanation of why the melt viscosity of sulphur is so low for its molecular weight.[9]

At temperatures below the crystalline melting point, polymeric sulphur is unstable and slowly reverts to crystalline (rhombic) cyclo-octasulphur. Because of the potential value of such a cheap inorganic elastomer, many attempts have been made to arrest this change. The addition of very small amounts of either phosphorus or arsenic to the molten polymer greatly reduces the rate of crystallization, but its potentially useful elastomeric properties alter dramatically with increasing amounts of these additives because the glass transition temperature rises steeply (see Chapter 3, p.49). Thorough extraction of the polymer with carbon disulphide to remove the small equilibrium concentration of cyclo-octasulphur always present after quenching also tends to delay the onset of crystallization, presumably by the removal of nuclei. Currell[10] found that the addition of olefines such as myrcene, limonene and cyclopentadiene also retard the crystallization of the free cyclo-octasulphur; similar effects were produced by admixture with polymeric polysulphides of the "Thiokol" type, but in this case the products slowly evolved hydrogen sulphide at room temperature, and also became hard and friable within a week to 10 days.

Polymeric Selenium

Selenium occurs in three crystalline forms. The two monoclinic varieties, known as α- and β-selenium, are obtained by precipitation from solution and consist of eight-membered rings corresponding to cyclo-octasulphur. Above 180°C both are transformed into hexagonal selenium, which consists of long chain molecules. Since the melting point of selenium is 217°C, there is no part of the liquid range in which the melt consists only of cyclic molecules, as there is in the case of sulphur. Eisenberg and Tobolsky,[11] using the theory outlined above for the equilibrium polymerization of sulphur, predicted that the transition to Se_8 rings should occur at 82°C, were it possible to maintain selenium molten at this temperature.

Like many organic polymers (such as polyethylene), polymeric selenium crystallizes into spherulitic structures characterized by radial lamellar growth. The polymer chain is oriented perpendicularly to the lamellae, which can vary in thickness from about 10–20 nm, suggesting chain folding in the crystal. Crystal[12] has shown that chain-folded selenium transforms into an extended-chain form above 125°C by a process of bond scission at the folds, followed by secondary crystallization and interlamellar bond formation.

The viscosity of molten selenium is about an order of magnitude less than that of sulphur at similar temperatures, even though the concentration of polymer in the melt is greater, indicating that the number-average degree of polymerization must be much lower. In fact the molecular weight of selenium at 220°C is only about 800,000, compared with an estimated 12,000,000 for sulphur at the same temperature. On rapid cooling, molten selenium is readily converted into a glass with a glass transition of 31°C. Above 70°C this glass readily devitrifies; the rates of nucleation and crystal growth have been measured at various temperatures by Borelius[13] and by Hillig.[14] The crystallization of amorphous selenium is considerably accelerated by minor amounts of the halogens, the alkali metals and thallium; all of which lower the melt viscosity, presumably by acting as chain-terminators. However, tellurium also has the same effect and its mode of action is not clear. Possibly the presence of tellurium atoms in the chains facilitates bond interchange.

Copolymers of sulphur and selenium have been studied by Tobolsky and Owen[15] and by Ward and Meyers,[16] The Raman spectra of sulphur-selenium mixtures show bands attributable to the presence of ring molecules containing both types of atom, such as S_4Se_4. This has a pronounced effect on the temperature-dependence of the polymerization reactions. Addition of selenium to sulphur markedly lowers the polymerization temperature of sulphur, because the bond energy of Se—S is lower than that of S—S, and in consequence the enthalpy of the propagation reaction

$$\sim\sim S\cdot + S_4Se_4 \rightarrow \sim\sim S\cdot Se_4\cdot S_4\cdot$$

is lower than in pure sulphur. Thus, while polymerization begins in pure sulphur at 159°C, polymerization of sulphur containing 10 atom% of selenium commences at 110°C.

Assuming that the mixed sulphur-selenium cyclic monomer has the composition Se_3S_5, Ward[17] has calculated the molecular composition of the copolymers formed. For instance, a melt containing 95 atoms of Se and 5 atoms of S per 100 atoms will form 57·3% by weight of a polymer with average composition $Se_{23}S$, the remainder of the melt being Se_3S_5 (2·7%) and Se_8 (40%). Although Raman spectroscopy suggests that the assumption of Se_3S_5 as the composition of the comonomer is an over-simplification, the calculated equilibrium composition of the polymer is not very sensitive to the exact composition assumed for the mixed monomer. Ward also showed that the number-average chain length of the Se—S mixed polymer increases as sulphur is added to the system.

Arsenic in small quantities also lowers the polymerization temperature of both sulphur and selenium; as little as 2 atom % of As lowers the polymerization temperature of sulphur from 159°C to 122°C, but larger quantities produce little further effect, apart from branching and cross-linking in the polymer (see Chapter 3, p.49).

Polymeric Tellurium

Unlike the other chalcogens, tellurium only occurs in one crystalline form which is hexagonal and consists of long-chain molecules in an extended-chain configuration. Tellurium has a high viscosity within 10° of its melting point, but the viscosity falls quickly at higher temperatures indicating a rapid decrease in molecular weight. The melt cannot be quenched to a glass, but an amorphous form of tellurium has been obtained by condensing the vapour on to a cold surface.[18] The resulting film crystallized readily at 30°C.

Selenium and tellurium are isomorphous and form copolymer crystals, and the effect of tellurium on the crystallization of polymeric selenium has been studied by Crystal,[19] who found that the thickness of the lamellar folds increased with increasing tellurium content. Scanning electron microscopy revealed interlamellar linkages, indicating the presence of polymer chains that extend over more than one crystal.

Linear Polyphosphates ("metaphosphates")

Substantially linear metal and ammonium polyphosphates of high molecular weight can be prepared by thermal dehydration of the corresponding metal or ammonium dihydrogen phosphate:

$$n\mathrm{MH_2PO_4} \rightarrow (\mathrm{MPO_3})_n + n\mathrm{H_2O}.$$

The type of polymer formed, its molecular weight, crystallinity and other properties depend not only upon the nature of the cations present, but also upon the conditions used for the dehydration. Since almost all polyphosphates other than those of the alkali metals and ammonium are insoluble in water, they can also be obtained by precipitation from a suitable alkali metal polyphosphate solution by addition of an appropriate metal salt. When produced by this method, many divalent metal polyphosphates are elastomeric, gummy materials that readily retain considerable amounts of water (up to 50%); but the corresponding anhydrous polymers are brittle, glassy or crystalline solids, which are insoluble in and not swollen by water.

Quaternary ammonium polyphosphates with at least one long-chain alkyl group attached to the nitrogen atom are like greases in consistency and some of them have been evaluated as extreme-pressure lubricants.

Alkali Metal Polyphosphates

The minimum temperatures that are required to bring about the polycondensation of the two common alkali metal dihydrogen orthophosphates to high molecular weight polymers are about 340°C for NaH_2PO_4 and 230°C for KH_2PO_4. Dehydration at lower temperatures gives low molecular weight oligomers, chiefly cyclic trimers and tetramers.

Sodium polyphosphate occurs in four varieties, of which three are crystalline and one is a glass. The three crystalline forms differ in the geometrical arrangement of the PO_4 tetrahedra along the chain; dehydration of sodium dihydrogen phosphate below 475°C gives Maddrell's salt, which consists of extended chains with a repeat distance in the crystal corresponding to three PO_4 units. Kurrol's salt, which is obtained by crystallization from molten sodium polyphosphate at 550–650°C, contains a helical chain with a repeat distance corresponding to four PO_4 units. The crystalline sodium polyphosphates are remarkable for the negligibly small rate at which they are dissolved by pure water; in the presence of even a small amount of a different alkali cation, however, they dissolve readily to form viscous solutions. Adding another alkali metal salt, however, necessarily introduces a foreign anion. A convenient way to obtain alkali polyphosphate solutions free from any other anion is to agitate the crystalline polymer with an aqueous suspension of an insoluble cation-exchange resin that has been exchanged with a different cation.[20]

Heating either of the crystalline forms of sodium polyphosphate above their melting point (627·6°C), or dehydrating sodium dihydrogen phosphate at 650°C, produces a molten polymer which cools to a glass. The glassy form of sodium polyphosphate, unlike the crystalline varieties, dissolves readily in water to give highly viscous solutions.

Potassium polyphosphate is formed by heating potassium dihydrogen phosphate at any temperature above 250°C until the theoretical loss in weight corresponding to the reaction

$$n\mathrm{KH_2PO_4} \rightarrow (\mathrm{KPO_3})_n + n\mathrm{H_2O}$$

has occurred. Like sodium polyphosphate, this polymer only dissolves very slowly in water, but dissolves readily in solutions of salts of other cations or by reaction with a suspension of a cation-exchange resin. Pfanstiel and Iler[21] found that the molecular weight of the polymer, as measured by its solution viscosity, was practically independent of the temperature used in the polycondensation up to 807°C, the melting point of the polymer; but the viscosity of its solution is very sensitive to the ratio of potassium to phosphorus in the starting material. With a small excess of phosphorus over the stoichiometric ratio of 1:1, a common occurrence in ordinary reagent quality potassium dihydrogen phosphate due to a trace of phosphoric acid as impurity, the resulting polymer is partially cross-linked and only swells in water to a gummy consistency without dissolving properly. With a small excess of potassium that can arise from the presence of dipotassium hydrogen phosphate as impurity, the polymer has a greatly reduced solution viscosity, but this is not due to a lower molecular weight. Pfanstiel and Iler showed that the lower viscosity in this case was attributable to the presence of potassium tripolyphosphate, which reduces the viscosity of the polymer in solution because it is a low molecular weight electrolyte. By extracting the polymer with water to remove the much more soluble tripolyphosphate before preparing the solution, its specific viscosity is restored almost to the value that is obtained for the polymer made from pure potassium dihydrogen phosphate. They concluded that dipotassium hydrogen phosphate does not react as a chain terminator in this polycondensation reaction as might be expected, but instead reacts independently to form pentapotassium tripolyphosphate:

$$KH_2PO_4 + 2K_2HPO_4 \rightarrow K_5P_3O_{10} + 2H_2O.$$

When potassium polyphosphate is heated above its melting point, the resulting liquid does not have the high viscosity typical of a molten polymer, and when rapidly chilled between steel plates it forms a translucent, only partially vitreous solid which dissolves readily in water to give solutions of very low specific viscosity. This shows a significant difference in behaviour from sodium polyphosphate, which forms a highly viscous melt that solidifies to a polymeric glass. The difference can be attributed to the much higher liquidus temperature (807°C) of the potassium salt. At this temperature there is a substantially higher concentration of comparatively low molecular weight polymer chains in the equilibrium mixture. If molten potassium polyphosphate is cooled slowly, however, so that the equilibrium molecular weight is established, a comparatively high molecular weight polymer crystallizes at about 775°C. Potassium polyphosphate of high molecular weight can evidently exists in only one form, which is crystalline. High molecular weight potassium polyphosphate, like sodium polyphosphate, is attacked so slowly by water as to be virtually insoluble, but it can be brought into solution by partial exchange of potassium for another alkali cation. If the polymer is left to stand in a solution of a lithium salt, for example, it swells and acquires a jelly-like consistency. This swollen gel will then dissolve readily in salt-free or distilled water. In this way it is possible to prepare solutions containing as much as 25% by weight of polymer, but such solutions are invariably less viscous than solutions of similar concentration that have been made with the aid of a cation-exchange resin. This is probably due to the surplus electrolyte, which causes the polymer chains to adopt a tightly coiled configuration. Potassium polyphosphate will also dissolve in hydrogen peroxide solutions to give solutions that are nearly as viscous as those prepared by ion-exchange.

The molecular weight of potassium polyphosphates can be estimated by end-group titration in aqueous solution, and on this basis an empirical relationship between molecular weight and specific viscosity has been established:

$$\log_{10} P = 2 \cdot 12 + 0 \cdot 61 \log_{10} \eta_{sp},$$

where P is the degree of polymerization (the average number of phosphate groups per molecule) and η_{sp} is the specific viscosity at 1% concentration by weight.

Aqueous solutions of alkali polyphosphates very slowly become less viscous on standing at room temperature, and more quickly if heated, because of hydrolytic degradation:

$$\sim \underset{\underset{O^-}{|}}{\overset{\overset{O}{\|}}{P}} - O - \underset{\underset{O^-}{|}}{\overset{\overset{O}{\|}}{P}} \sim + H_2O \rightarrow 2 \sim \underset{\underset{O^-}{|}}{\overset{\overset{O}{\|}}{P}} - OH$$

The rate of the degradation reaction has been determined by viscosity measurements over the temperature range 25–90°C, and was found to be first-order with an activation energy of $104 \pm 8 \text{kJ mol}^{-1}$.

Polyvalent metal polyphosphates can be obtained by precipitation from a solution of sodium or potassium polyphosphates. If enough of a solution containing polyvalent cations is added to an alkali metal polyphosphate solution, a precipitate is formed that may either be particulate, as in the case of barium and strontium polyphosphates, or a gummy coacervate, as in the case of cobalt and magnesium polyphosphates. When the polymer coagulates into a gum, the replacement of cations is generally far from complete and the product can be redissolved in pure water. Successive precipitations will eventually yield an insoluble product which, when freshly prepared, contains up to 50% of water. These materials slowly dry out, passing through a partially hydrated condition in which they behave as uncross-linked rubbers; then with further drying, gradually becoming stiffer and more brittle, until in the anhydrous state they become hard, brittle glasses. This change is irreversible and the dried glass cannot be reconverted to the rubber by any kind of treatment with water. No means of inhibiting or delaying the hardening process has yet been found; magnesium polyphosphate even hardens slowly under water, showing that a structural change to a more stable configuration must be involved, and it is not sufficient merely to prevent evaporation of water from the gum.[22] Replacement of water by other liquids such as glycerol has been tried without success.

Many divalent metal polyphosphates have also been prepared as glassy polymers by dehydration of the corresponding metal dihydrogen phosphates. When the glass transition temperatures of the amorphous forms of a number of polyphosphates are plotted against their oxygen density (the number of g atoms of oxygen per litre of polymer), which is a convenient way of measuring the relative packing density of the polymer molecules in an oxide-linked glass, two approximately parallel curves separated by about 200°C are obtained.[23] The glass transition temperatures of the alkali metal polyphosphates lie some 200°C lower than those of the alkaline earth metal polyphosphates at the same packing density of polymer molecules; while for both sets of cations, the glass transition temperature rises with increasing packing density, that is with decreasing size of the cation (Fig. 2.1).

The effects of cation size and charge on the glass transition temperature of polyphosphates have also been studied by Eisenberg *et al.*[24] who found that for many cations, the glass transition temperature increases in direct proportion to the ratio of ionic charge to the sum of the cation and oxygen radii, i.e.

$$Tg = KQ/R,$$

where Q is the charge on the cation and R is the sum of the cation and oxygen radii. Some values of Q/R for a number of metals are given in Table IV, which includes a few additional results that do not fit Eisenberg's relationship (Fig. 2.2). In particular, the glass transition temperature of silver and lead^{II} polyphosphates are very much lower than would be expected from their charge:radius ratio. A possible explanation of this is that the structures of these and some other metal polyphosphates are different from those of the simple alkali and alkaline earth metal polyphosphates. In a study of linear polyphosphates containing both lead and potassium ions by Raman spectroscopy,[25] the present author found that the ratio of PO_4 to P—O—P units deviated from the expected value of unity, the ratio increasing with increasing lead content. Both this and the unusually low glass transition temperature of lead polyphosphate glass may be due to the incorporation of lead atoms into the polymer chain to form P—O—Pb units; this type of structural change is known to occur in borate and silicate glasses containing high proportions of lead oxide.

Quaternary ammonium polyphosphates can best be obtained by precipitation from a solution of a soluble alkali polyphosphate, apart from tetramethylammonium polyphosphate, which is soluble in water and must therefore be prepared by ion-exchange. Quaternary ammonium ions containing one or more long-chain alkyl radicals are insoluble in water and are like greases in consistency. These polymers are soluble in chloroform, toluene and cellosolve, and are useful as boundary lubricants.[26]

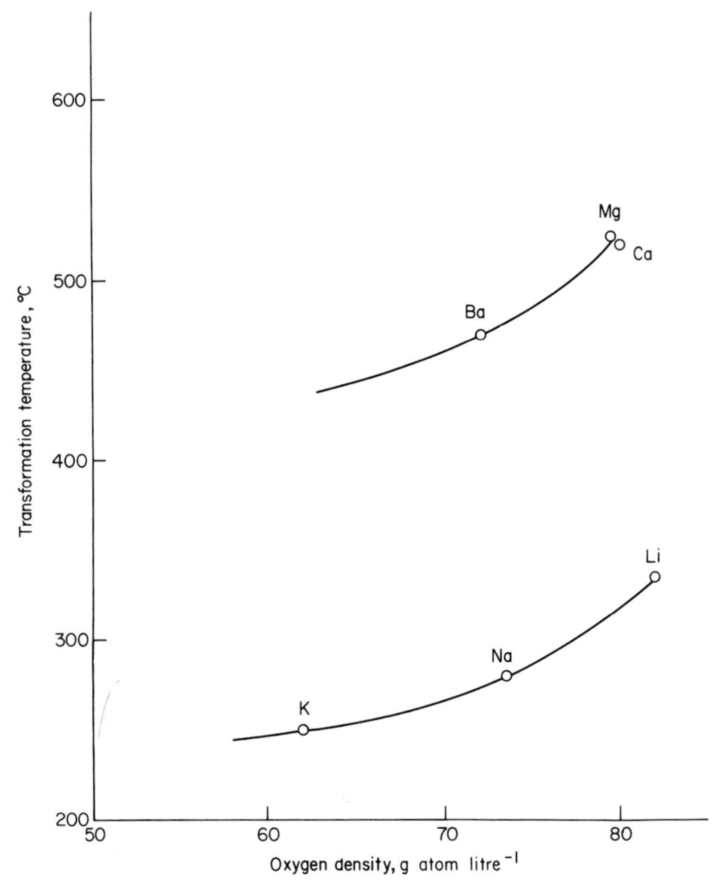

Fig. 2.1. Glass transition temperatures of alkali and alkaline earth linear polyphosphates plotted against oxygen density. (From Ray, N. H. (1974), *J. Non-cryst. Solids,* **15**, 428. With permission.)

Linear polyphosphates containing mixed cations, which may be regarded as copolymers, are easily prepared by treatment of an aqueous suspension of high molecular weight potassium polyphosphate with a cation exchange resin carrying a mixture of cations. For example, potassium polyphosphate treated with a mixture of the tetramethylammonium, magnesium and calcium salts of "Nalcite" HCR ion-exchange resin[27] gave a viscous solution that dried out to give strong, pliable, non-hygroscopic films. However, these films became brittle on prolonged drying and remained water-soluble. Mixed cetyl trimethylammon-

Table IV
Melting and glass transition temperatures of some metaphosphate polymers, $(MPO_3)_n$

Cation	Cation radius nm	Charge ÷ sum of cation + oxygen radii	Glass transition temp., °C.	Crystalline melting point, °C
Li	0·60	0·50	335	
Na	0·95	0·42	280	628
K	1·30	0·366	243	807
Mg	0·66	0·97	525	
Ca	0·99	0·84	520	975
Sr	1·13	0·79	485	
Ba	1·35	0·73	470	
Zn	0·74	0·93	520	
Cd	0·97	0·84	450	
Pb	1·20	0·76	315	667
Ag	1·26	0·37	115	

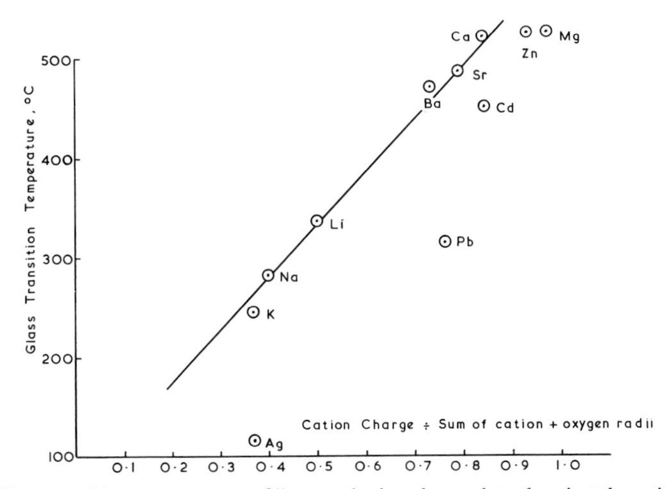

Fig. 2.2. Glass transition temperatures of linear polyphosphates plotted against the ratio of cation charge to sum of cation and oxygen radii.

ium-magnesium polyphosphates gave leathery products and cetyl trimethylammonium-barium polyphosphates containing about one Ba^{++} ion to four alkylamonium ions gave flexible films that could be cold-drawn; the inclusion of 3·3 Al atoms per 100 phosphate units gave a firm solid that extended 20% before fracture.

Polyphosphazenes

The polyphosphazenes are a family of polymers which ideally have the linear structure

$$(-PR_2=N-)_n$$

where R can be halogen, pseudohalogen, or an organic group such as alkyl, aryl, alkoxy or aryloxy. Polydichlorophosphazene has been known since 1897,[28] and because of its rubbery nature, unusual in a wholly inorganic material, it has attracted the attention of polymer scientists ever since. The exceptional thermal stability of some of the low molecular weight cyclic phosphazenes, such as hexa(trifluoroethoxy)cyclotriphosphazene, $[(CF_3CH_2O)_2PN]_3$, which can be distilled at 500°C without decomposition, has also stimulated many investigations of differently substituted phosphazenes with the object of discovering new heat-resistant flexible materials.

Preparation

Polyphosphazenes can be prepared in three ways.

(a) The reaction of ammonium chloride with phosphorus pentachloride, usually in tetrachloroethane, yields a mixture of products containing cyclic oligomers (mainly trimer and tetramer) together with a series of linear polymers with degrees of polymerization up to about 20, in proportions that vary with reaction conditions. The polymer produced in this reaction is apparently terminated by hydrogen and chlorine atoms,

$$Cl—(—PCl_2=N—)_n—H,$$

so that chain extension can be effected by reaction with a limited amount of water:

$$Cl—(—PCl_2=N—)_n—PCl_2=NH+H_2O$$
$$\rightarrow Cl—(PCl_2=N)_n—[PCl(NH)OPCl_2=N—]_m—H+2HCl.$$

(b) By ring-opening polymerization of the cyclic oligomers (generally the trimer). This was the method originally used by Stokes, but the only cyclic phosphazenes that give high polymers in this reaction are the halides and pseudohalides, and the polymers formed from then very readily become cross-linked. When pure hexachlorocyclotriphosphazene is heated in an evacuated, sealed tube at 250°C, it is slowly converted into linear polymer (60% in 56 h); but if it is heated for prolonged periods in an attempt to increase the yield, or at higher temperatures (above 300°C), or in the presence of catalysts such as benzoic acid, the product is a cross-linked polymer that is insoluble in benzene. Radiation-induced polymerization of the trimer takes place at much lower temperatures.

(c) Polyphosphazenes with a variety of substituents can be obtained by substitution of the chlorine atoms in polydichlorophosphazene. For this method to succeed, the chloropolymer must be linear so that it is soluble; early attempts to substitute the chlorine atoms in polydichlorophosphazene with other groups failed because the polymers used were cross linked and insoluble, so that only partial substitution of chlorine atoms was achieved. Cyclic trimers with substituents other than halogen or pseudo-halogen will not undergo ring-opening polymerization, and it was not until 1965 when Allcock *et al.* showed how to isolate a soluble, high molecular weight polydichlorophosphazene[29] that it became possible to synthesize phosphazene polymers with hydrolytically stable substituents. Complete substitution of the chlorine atoms in linear polydichlorophosphazene can be brought about by reaction with sodium alkoxides in tetrahydrofuran, and a variety of substituents have been incorporated in this manner, among which perfluoroalkoxy groups are the most important because they yield some of the more interesting and potentially useful products.

Substitution of the chlorine atoms can also be effected by reaction with amines; aniline and ethylamine react to give complete replacement, but ammonia and methylamine give cross-linked and insoluble products.[32] Because of steric hindrance, diethylamine will replace only half of the chlorine atoms, and consequently polymers with two different substituents can be made by a two-stage process involving a preliminary reaction with diethylamine, followed by reaction of the partially substituted product with either a primary amine or a sodium alkoxide.

It has been reported that a fully phenylated polymer can be made by reaction of polydichlorophosphazene with phenyl lithium, but Allen and co-workers were unable to repeat this.[30]

Properties

When freshly prepared the halogen-substituted polyphosphazenes are soluble in benzene and chloroform but insoluble in petroleum; the latter solvent can therefore be used to extract residual cyclic material from the polymer. On standing, either in solution or in the solid state, the halogen-substituted polymers undergo spontaneous cross-linking so that their solutions slowly gel. This reaction is accelerated by heating and by the addition of acid-acceptors, such as triethylamine. The polydihalogenophosphazenes are also hydrolytically unstable, reacting with atmospheric moisture to become brittle, cross-linked network polymers. The alkyl- and alkoxy-substituted polymers, on the other hand, are stable to hydrolysis and do not undergo spontaneous cross-linking.

For this reason, the alkoxy-substituted polyphosphazenes are easier to handle and have been more completely characterized. Cross-linked polydichlorophosphazene is an amorphous rubbery solid, but substitution of the chlorine atoms with other groups gives rise to a variety of substances with widely differing properties. Poly-dimethoxy- and poly-diethoxyphosphazenes are colourless, transparent, film-forming thermoplastics which unfortunately are thermally unstable, slowly reverting to the corresponding cyclic oligomers at temperatures above 100°C.

Trifluoroethoxy groups are readily introduced into the polymer by running a benzene solution of freshly prepared high molecular weight polydichlorophosphazene into a well stirred solution of sodium trifluoroethoxide in tetrahydrofuran, and precipitating the polymer into petroleum. Repeated extraction with water (to remove sodium chloride) and petroleum (to remove cyclic material) yields a product with a number average molecular weight of $1{\cdot}7 \times 10^6$ and an intrinsic viscosity of $2{\cdot}5$ dl g^{-1}. This can be fabricated into colourless, flexible films with a density of $1{\cdot}7$ g cm^{-3}, and can also be spun into fibres from solution. The glass transition temperature is −66°C and the polymer is crystalline up to the melting temperature of 242°C. It is soluble in acetone, tetrahydrofuran, ethyl acetate, dimethyl ether and methyl ethyl ketone, but insoluble in water, acetic acid, alcohols, dioxan and paraffins. It does not burn in air, but begins to decompose at 150°C, and degrades rapidly at 200°C. Misleading claims have been made for its thermal stability based on thermogravimetric analysis with rapid heating rates.[30, 31] Poly-(trifluoroethoxy-heptafluoro-butoxy) phosphazene, a copolymer of $NP(OCH_2CF_3)_2$ and $NP(OCH_2C_3F_7)_2$, is a rubber with a glass transition temperature at −77°C. This material softens at 175°C and decomposes at 260°C. Polyphosphazenes with fluorocarbon groups attached directly to phosphorus, like poly-bis-trifluoromethyl-phosphazene $[(CF_3)_{n2}PN]$, are white, amorphous, insoluble solids which decompose without softening appreciably at about 300°C, and which therefore cannot be fabricated; these materials are not elastomers.

Allen and co-workers[33] found that the aryloxy-substituted polyphosphazenes are more stable than the alkoxy derivatives; the latter were all completely vaporized below 550°C, whereas the aryloxy derivatives lost only 50% of their weight at 600°C. This difference appears to be due to a different mode of decomposition. The volatile product evolved from poly-di-*p*-phenylphenoxyphosphazene on heating at 600°C included *p*-phenylphenol, 4,4-diphenyldiphenylether and hexa-*p*-phenylphenoxy cyclotriphosphazene, suggesting that in the aromatic-substituted polymers, the side groups react to form cross-links below the temperature at which the main chain depolymerizes. This does not appear to occur with the alkoxy derivatives.

From the results of light-scattering measurements and osmometry on solutions of variously substituted polyphosphazenes, Allen, Lewis and Todd concluded that all their polymers had branched chains and broad molecular weight distributions.

X-ray examination of these polymers show patterns typical of semi-crystalline polymers. The chain repeat distance of 0·49 nm is hardly affected by the type or size of the substituents on the phosphorus atoms,[33, 34, 35] indicating that even quite bulky side-groups are capable of fitting into the chain without undue steric hindrance.

The *p*-phenylphenoxy-substituted polymer can be spun into fibres from the melt at 225°C, although there is some unavoidable degradation of the polymer under these conditions. The Young's modulus of the fibre is 2·14 $GN\,m^{-2}$, which is similar to that of nylon, and films cast from a tetrahydrofuran solution are rather like films of polyethylene.

The polyphosphazenes have a much lower electrical conductivity than the analogous carbon-based polyenes, suggesting that the π bonds are not delocalized, as is the case with conjugated double bonds in a carbon chain.

Table V.
Properties of some substituted polyphosphazenes, $(NPR_2)_x$

Substituent *R*	*Tg* °C	T_m °C	Temperature for 10% weight loss[a]
Cl	−64		410
F	−90		
OCH_3	−76		
OC_2H_5	−84		
OCH_2CF_3	−70	83, 240	410
$OCH_2C_2F_5$	−73		
$OCH_2C_3F_7$	−70		
$OCH_2(CF_2)_6CF_3$	−40		425
OCH_2CF_3 & $OCH_2C_3F_7$[b]	−77	175	
OC_6H_5	− 8		
OC_6H_4Cl(p)	−12	165, 405	
$OC_6H_4CF_3$(m)	−35	330	380
OC_6H_4F(p)	−14		385
$OC_6H_3Cl_2$(2,4)	+ 2	195	440
$OC_{10}H_7(\beta)$	47	160	475
$OC_6H_4C_6H_5$(p)	43	160, 398	390
NHC_2H_5	30		
NHC_6H_5	91		

[a] Heating at 10°C min^{-1} in dry nitrogen.
[b] Approximately 1:1 random copolymer.

The properties of a number of polyphosphazenes with different substituents are summarized in Table V; some of these polymers have two first-order transitions, but only the upper one leads to the total disappearance of crystallinity.

Polymeric Sulphur Nitride

Polymeric sulphur nitride, $(SN)_x$, was first prepared by Burt in 1910[36] by passing the vapour of cyclic tetrasulphur tetranitride over silver gauze or silica wool at 100–300°C. The polymer was deposited on a cooled glass surface in the form of a thin film, which was at first blue by transmitted light, and which subsequently became opaque and acquired a bronze metallic lustre as it increased in thickness. In 1971 Boudeulle and others[37] reported that the primary product that is formed when tetrasulphur tetranitride is passed over a heated silver surface is the very unstable dimeric compound S_2N_2, which polymerizes rapidly and spontaneously at low temperatures to give pseudo-single crystals of polymer, composed of layers of fibre-like crystals of $(SN)_x$. In 1973, Labes and co-workers[38] synthesized crystalline polymeric sulphur nitride in the form of bundles of fibres, and found that the material possesses metallic conductivity in a direction parallel to the length of the fibres. The conductivity increases with decreasing temperature, and in early preparations the specific resistivity varied from specimen to specimen, probably (in view of later work) because of the presence of variable amounts of impurity.

Subsequently, MacDiarmid and his team[39] prepared analytically pure polymeric sulphur nitride by pumping tetrasulphur tetranitride vapour over silver wool at 220°C and condensing the disulphur dinitride formed on to a liquid nitrogen-cooled surface. The disulphur dinitride was then sublimed into an ice-cooled trap and slowly allowed to warm up to room temperature. The colourless monoclinic crystals of S_2N_2 rapidly turned blue-black and became paramagnetic. After several hours they became opaque and acquired a bright, lustrous golden colour; at this stage the product still contained some unpolymerized S_2N_2 which was finally removed by pumping out at 75°C. The polymer was shown to be free from S_4N_4 and S_2N_2 by X-ray diffraction photographs.

Electron microscopy of this highly anisotropic crystalline polymer shows that it consists of layers of fibres stacked parallel to the long axis of the crystal. When the crystals are separated mechanically, long strands of polymer are visible at their junctions. Quite apart from its metallic-type electrical conductivity, this interesting polymer shows some of the physical properties of a ductile metal. It is relatively soft and malleable, and the crystals can be flattened under pressure into thin, lustrous sheets. After some months in the air, it becomes coated with a whitish powdery deposit, and a white solid is also formed when the polymer is left for six days in contact with water, and the water is then evaporated. The polymer decomposes in a vacuum at 140°C to give sulphur, nitrogen, and some other unidentified material; but by sublimation *in vacuo* at about 140°C it is possible to obtain coherent polycrystalline films on a surface cooled to 0°C.

This polymer consists of a nearly planar chain of alternating S and N atoms, in which all the S—N bonds are of approximately the same length (0·16 mm) and the bond angles are S—N—S, 119·9°; N—S—N, 106·2° (Fig. 2.3). The

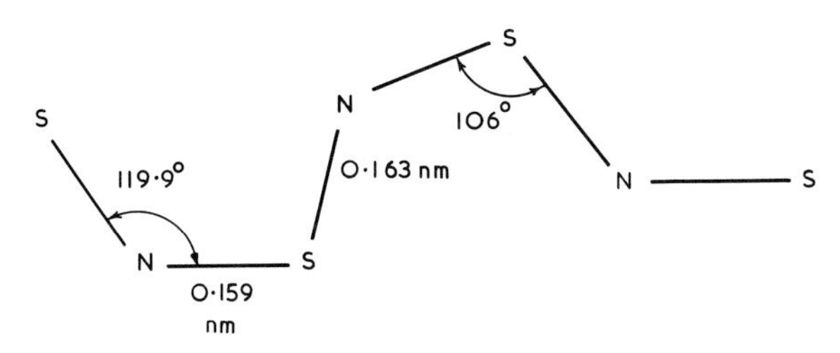

Fig. 2.3. Structure of the polymer chain in crystalline polymeric sulphur nitride.

unit cell contains four monomer units, with crystal space group P $2_1/c$ and cell dimensions $a = 0{\cdot}4153$, $b = 0{\cdot}4439$, $c = 0{\cdot}7637$ nm; $\beta = 109{\cdot}7°$; its true density is 2·30 g cm^{-3}. MacDiarmid concluded that there must be weak interchain forces between the polymer molecules from the observation that some of the S—S distances in the crystals are less than the sum of the unperturbed van der Waals radii of the two atoms. This effect is probably responsible for the material behaving as a two-dimensional, rather than simply as a one-dimensional conductor. The conductivity of polymeric sulphur nitride at room temperature was found to be in the range 1200 to 3700 S cm^{-1}, and was increased by 50 to 200 times when the temperature was lowered to 4·2 K. This remarkable polymer therefore has a similar conductivity to mercury and bismuth. Recently, Greene[(40)] found that it becomes superconducting at 0·26 K; this is therefore the first example of a non-metallic polymeric superconductor.

Polycarboranes

The polycarboranes are mainly linear polymers built up from meta- or paracarborane units (1,7- or 1,12-dicarbaclosododecaborane) linked through a variety of hetero-atom bridges, of which the most important are —Si—O—Si— and —P—O—P—.

Orthocarborane (Fig. 2.4a) is obtained by addition of acetylene to decaborane[41,42] to form a closed icosahedral molecule $C_2B_{10}H_{12}$ with a diameter of about 0·4 nm, in which the two carbon atoms are adjacent. On heating this compound to 475°C it undergoes an intramolecular rearrangement to give the isomeric metacarborane, in which the carbon atoms are separated by one boron atom (Fig. 2.4b); while on further heating above 630°C, this intermediate is itself isomerized to paracarborane (Fig. 2.4c) in which the carbon atoms lie at opposite vertices of the cage.[43,44]

In these compounds the hydrogen atoms attached to carbon are acidic and can be replaced by lithium:

$$C_2B_{10}H_{12} + 2\ Li \rightarrow C_2B_{10}H_{10}Li_2 + H_2.$$

The dilithium salts react readily with halogen compounds such as phosgene, dimethyldichlorosilane and phosphorus trichloride, to give disubstituted carboranes with reactive groups.[45] In the case of dilithio-orthocarborane, the main products from this type of reaction were cyclic compounds:

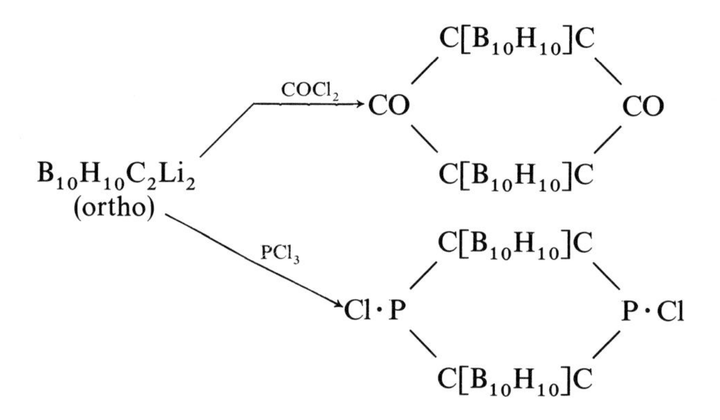

In similar reactions, the lithium salts of meta- and paracarboranes gave rise to linear polymers, but with carbon or phosphorus as linking atoms between the carborane units, the products were only low molecular weight oligomers containing no more than about five repeat units. Reaction with dimethyl tin dichloride in xylene or decalin, however, gave crystalline polymers of reasonable chain length (20–30 repeat units) with softening points about 240–250°C:[46]

$$Li\cdot C\ B_{10}H_{10}\ C\cdot Li + Me_2SnCl_2 \xrightarrow{-2\ LiCl} —C\ B_{10}H_{10}\ C\cdot SnMe_2—$$

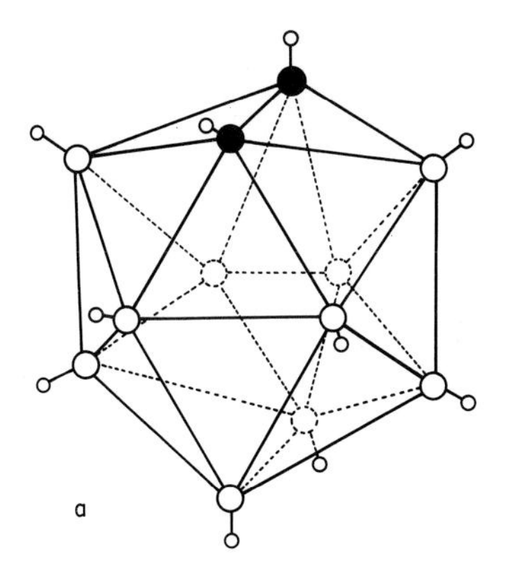

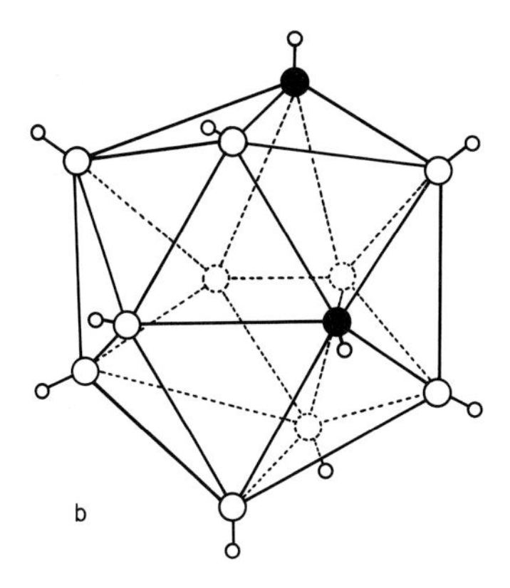

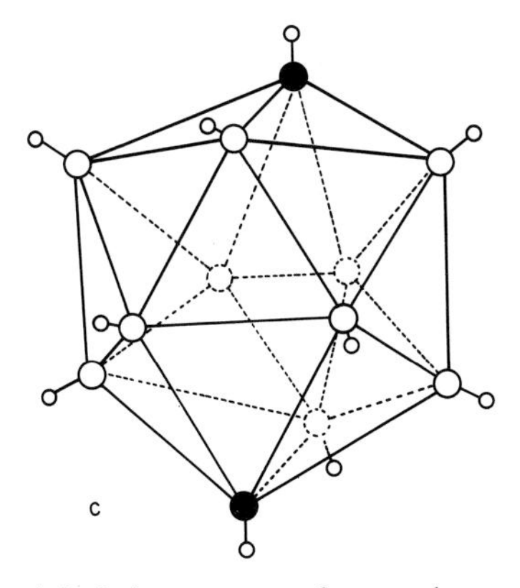

Fig. 2.4. Ortho-, meta-, and paracarboranes.

Similar polymers were also obtained by reaction with dimethyl germanium dichloride,[47] but some of the most interesting carborane polymers are those in which the carborane units are linked through short polysiloxane bridges. Reaction of dilithio-metacarborane with an excess of dimethyldichlorosilane gives 1,7-bis(dimethylchlorosilyl)-1,7-dicarbaclosododecarborane, and the corresponding methoxy derivative is obtained by reaction with methanol:

$$\mathrm{Li\cdot C[B_{10}H_{10}]C\cdot Li + Me_2SiCl_2 \longrightarrow Cl\cdot SiMe_2\cdot C[B_{10}H_{10}]C\cdot SiMe_2Cl}$$

$$\downarrow \mathrm{MeOH}$$

$$\mathrm{MeO\cdot SiMe_2\cdot C[B_{10}H_{10}]C\cdot SiMe_2\cdot OMe}$$

Polycondensation of an equimolar mixture of these two compounds at 190°C in the presence of ferric chloride results in quantitative evolution of methyl chloride and the formation of a crystalline polymer with a glass transition temperature of 25°C and a crystalline melting point of 240°C.[48] This is the first member of a series of polymers with the general formula

$$\mathrm{—SiMe_2\cdot C\,[B_{10}H_{10}]\,C\cdot\,[SiMe_2O\,]_n—},$$

where n can be from 1 to 4, to which the generic name "Dexsil" has been given by the Olin Research Centre, where these materials were originally developed. The second member of the series ($n = 2$) is obtained by reaction of the dimethoxy derivative with dimethyldichlorosilane, and the third member with tetramethyldichlorodisiloxane. Apart from the first member of the series, all these polymers are partially crystalline elastomers, with glass transition temperatures that decrease with increasing numbers of silicon atoms in the linking groups. In order to make it possible to cure these elastomers into useful rubbers, a small proportion of vinyl groups are incorporated by adding a minor amount of 1-vinyl-2-methyldichloro-1,2-dicarbaclosododecaborane to the polycondensation reaction. The resulting elastomer can be cross-linked with peroxide initiators, and the properties of the resulting rubber can be still further improved by compounding with finely divided silica fillers such as "Cab–O–Sil".[49] Rubbers of this composition are stable in air up to 260°C, but oxidative decomposition begins at 290°C. In nitrogen, these polymers are more stable at 300°C than silicone rubbers.

Polymers have also been prepared from meta- and paracarborane units linked through tin, germanium and sulphur atoms; reaction of dilithioparacarborane with dimethyl tin dichloride gave an insoluble polymer that softened only above 420°C but decomposed at 450°C. The corresponding reaction with dimethyl germanium dichloride gave polymers with somewhat similar properties, and mixed tin- and germanium-linked polymers have been described. Chlorination of 1,7-dimercaptocarborane[50] gave the bis(sulphenylchloride) which reacted with dilithiocarborane to give a linear polymer with an average molecular weight of 5400, corresponding to 30 repeat units. When metacarborane units are linked in this way, the polymer melts at 231°C, but with paracarborane units the crystalline polymer softens only above 420°C.

While the carborane polymers are superior in high-temperature performance to all but a very few organic polymers, they all contain appreciable quantities of hydrogen, and consequently their oxidation is exothermic, like that of organic polymers. The chief limitation on their use is therefore set not simply by thermal stability, but by the temperature at which the rate of oxidation in air becomes appreciable, which for most polycarboranes is about 300–310°C.

References

1. Gee, G. (1952). *Trans. Faraday Soc.* **48**, 515.
2. Fairbrother, F., Gee, G. and Merrall, G. T. (1955). *J. Polymer Sci.* **16**, 459.
3. Tobolsky, A. V. and Eisenberg, A. (1959). *J. Amer. Chem. Soc.* **81**, 780.
4. Schmidt, M. and Wilhelm, E. (1966). *Angew. Chem.* **78**, 1020.
5. Van Wazer, J. R. (1970). *Inorg. Macromol. Rev.* **1**, 94.
6. Eisenberg, A. (1969). *Macromolecules,* **2**, 44.
7. Tobolsky, A. V. (1966). *J. Polymer Sci.* Part C, **12**, 71.
8. Eisenberg, A. and Teter, L. A. (1967). *J. Phys. Chem.* **71**, 2332.
9. Kende, I., Pickering, T. L. and Tobolsky, A. V. (1975). *J. Amer. Chem. Soc.* **87**, 5582.
10. Currell, B. R., Williams, A. J., Mooney, A. J., and Nash, B. J. (1975). Advances in Chemistry Series; *New Uses of Sulphur Symposium 1974,* **1**, 1.
11. Eisenberg, A. and Tobolsky, A. V. (1960). *J. Polymer Sci.* **46**, 19.
12. Crystal, R. G. (1970). *J. Polymer Sci.* A2, **8**, 2153.
13. Borelius, G., Philstrand, F., Anderson, J., Gullberg, K., Glansholm, D., and Ekedahl, M. (1949). *Ark. Fys.* **1**, 305.
14. Hillig, W. B. (1956). *J. Phys. Chem.* **60**, 56.
15. Tobolsky, A. V. and Owen, G. D. T. (1962). *J. Polymer Sci.* **59**, 329.
16. Ward, A. T. and Myers, M. P. (1969). *J. Phys. Chem.* **73**, 1374.
17. Ward, A. T. (1968). *J. Phys. Chem.* **72**, 4133.
18. Suhrmann, R. and Berndt, W. (1940). *Z. Phys.* **115,** 17.
19. Crystal, R. G. (1972). *Polymer Preprints* **13** (2), 800.
20. Iler, R. K. (1951). U.S. Patent 2557109.
21. Pfanstiel, R. and Iler, R. K. (1952). *J. Amer. Chem. Soc.* **74,** 6059.
22. Pfanstiel, R. and Iler, R. K. (1956). *J. Amer. Chem. Soc.* **78**, 5510.
23. Ray, N. H. (1974). *J. Non-crys. Solids,* **15**, 423.
24. Eisenberg, A., Farb, H. and Cool, L. G. (1966). *J. Polymer Sci.* A2, **4**, 855.
25. Ray, N. H. (1975). *Glass Technology,* **16**, 107.
26. British Patent Applications, Nos. 26265/54 (1954), 7174/55, 5402/55, 11384/55, 26959/55 (1955).

27. Iler, R. K. (1952). *J. Phys. Chem.* **56**, 1086.
28. Stokes, H. N. (1897). *Amer. Chem. J.* **19,** 782.
29. Allcock, H. R., Kugel, R. L. and Valan, K. J. (1966). *Inorg. Chem.* **5**, 1709.
30. Allen, G., Lewis, C. J. and Todd, S. M. (1970). *Polymer,* **11**, 31.
31. Rose, S. H. (1968). *J. Polymer Sci.* B, **6**, 837.
32. Allcock, H. R. and Kugel, R. L. (1966). *Inorg. Chem.* **5**, 1716.
33. Allen, G., Lewis, C. J. and Todd, S. M. (1970). *Polymer* **11**, 44.
34. Giglio, E., Pompa, F. and Ripamonti, A. (1962). *J. Polymer Sci.* **59**, 293.
35. Allcock, H. R., Konopski, G. F., Kugel, R. L. and Stroh, E. G. (1970). *Chem. Commun.* 985.
36. Burt, F. P. (1910). *J. Chem. Soc.* 1171.
37. Boudeulle, M., Douillard, A., Michel, P., and Vallet, G. (1971). *Comptes rend.,* Ser. C, **272**, 2137.
38. Walatka Jr., V. V., Labes, M. M., and Perlstein, J. H. (1973). *Phys. Rev. Letters,* **31**, 1139.
39. MacDiarmid, A. G., Mikulski, C. M., Russo, P. J., Saran, M. S., Garito, A. F. and Heeger, A. J. (1975). *Chem. Commun. 476*; *J. Amer. Chem. Soc.* **97**, 6358.
40. Green, R. L., Street, G. B. and Suter, L. J. (1975). *Phys. Rev. Letters,* **34**, 577.
41. Heying, T. L., Ager, J. W., Clark, S. L., Mangold, D. J., Goldstein, H. L., Hillman, M., Polak, R. J. and Szymanski, J. W. (1963). *Inorg. Chem.* **2**, 1089.
42. Grafstein, D., Bobinski, J., Dvorak, J., Smith, H., Schwartz, N., Cohen, M. S. and Fein, M. M. (1963). *Inorg. Chem.* **2**, 1120.
43. Schroeder, H. A. and Vickers, G. D. (1963). *Inorg. Chem.* **2**, 1317.
44. Papetti, S. and Heying, T. L. (1964). *J. Amer. Chem. Soc.* **86**, 2995.
45. Papetti, S. and Heying, T. L. (1964). *Inorg. Chem.* **3**, 1448.
46. Schroeder, H. A., Papetti, S., Alexander, R. P., Sieckhaus, J. F. and Heying, T. L. (1969). *Inorg. Chem.* **8**, 2444.
47. Bresadola, S., Rossetto, F. and Tagliavini, G. (1968). *European Polymer J.* **4**, 75.
48. Papetti, S., Schaeffer, B. B., Gray, A. P. and Heying, T. L. (1966).
49. Schroeder, H. A. (1969). *Rubber Age,* **101** (2), 58.
50. Smith, H. D., Obenland, C. O. and Papetti, S. (1966). *Inorg. Chem.* **5**, 1013.

3

Three-connective Network Polymers

Chalcogenide Glasses

The chalcogenide glasses are a family of amorphous cross-linked polymers formed by compounds of the chalcogens sulphur, selenium and tellurium, with one or more of the polyvalent elements antimony, arsenic, bismuth, cadmium, gallium, germanium, indium, lead, mercury, phosphorus, silicon, tin and thallium. The halogens (except fluorine) may also occur in a chalcogenide glass as chain-terminating groups. These polymers can have glass transition temperatures as low as $-17°C$ or as high as $400°C$; they are all deeply coloured and many are transparent only in the infra-red. They are fairly stable to acids but are attacked by concentrated alkalies, and most chalcogenide polymers begin to oxidize in air at about $300°C$.

The high melt viscosity of these compounds leaves no room for doubt that they are polymers; yet some members of this group can be distilled *in vacuo* without decomposition, a behaviour that is normally quite untypical of polymeric substances. Their volatilization almost certainly involves reversible dissociation of the polymer network into small fragments which recombine spontaneously when the vapour condenses.

The first published description of a chalcogenide glass by Schultz–Sellack appeared in 1870,[1] but it was not until after the Second World War that detailed systematic work on these materials really began with the work of Frerichs[2] and Fraser.[3] Frerichs made a pure arsenic sulphide glass by distilling As_2S_3 in a current of hydrogen sulphide and showed it to be a good infra-red transmitting material. Substantial quantities of optical quality arsenic sulphide glass were manufactured during the decade following the war, mainly for use as infra-red windows. The unusual electrical properties of some of the chalcogenide glasses were first reported by Pearson[4] in 1962, and the so-called Ovshinsky effect (actually discovered by Pearson) became a popular subject for research in 1968; but the poor reliability and variable characteristics of experimental switching and memory devices made from the chalcogenide glasses have inhibited any subsequent commercial development.

Binary Chalcogenide Glasses

Homogeneous glass-forming polymers exist over a wide range of compositions in the systems listed in Table VI, and some of these systems also include crystalline compounds with polymeric structures. The best known of the binary chalcogenide polymers is arsenic trisulphide, the only material of this type to have been manufactured on a commercial scale.

Table VI
Binary chalcogenide polymers

System A—B	Glass-forming compositions atom % A	Crystalline polymers	
		compound	atom % A
As—S	5–44	As_2S_3	40
As—Se	0–39		
As—Te	30–50		
Sb—S		Sb_2S_3	40
Sb—Se		Sb_2S_3	40
Ge—S	0–33		
Ge—Se	0–25		
Si—S	25–50	SiS_2	33·3

Multicomponent Chalcogenide Glasses

A much larger number of three-component chalcogenide glass-forming systems is known, and in many of them the glass-forming composition range is more extensive than it is in any of the constituent pairs. For example, the glass-forming regions in the systems Si—P—Te, Si—Sb—Te, Ge—P—Se and Si—Sb—S do not intercept the two-component boundaries of the phase diagrams. Four- and five-component chalcogenide polymers are very numerous, typical examples being As—Ge—Si—Te, As—Ge—P—Se and As—Cd—Ge—S. There is also one naturally occurring three-component chalcogenide glass of approximate composition $As_2Pb_5S_8$.

Preparation

Chalcogenide polymers are generally made by fusion of the elements under conditions selected to minimize both oxidation and loss by volatilization of the components. In the laboratory, evacuated and sealed silica tubes are commonly used for reaction vessels, but these have the disadvantage that it is difficult to ensure homogeneity of the contents. Because of the high viscosity of the melt, simple rotation or overturning of the tubes is usually inadequate to mix the contents thoroughly. On a somewhat larger scale, stainless steel autoclaves with mechanical stirrers have generally been used with success. To ensure complete reaction the batch must be heated to temperatures between 500° and 1100°C, for periods ranging from 5–24h. Glasses that are near to a stoichiometric composition, such as As_2S_3 and SiS_2, can be purified by vacuum distillation, and thin films of these substances can be prepared by vacuum evaporation either of the preformed glass or of a mixture of the components.

Structure

X-ray studies of vitreous arsenic sulphide, selenide and telluride by Vaipolin and Porai–Koshits[5] indicated that these substances have structures built up of puckered layers. Recent structural investigations of arsenic selenide and arsenic telluride glasses have shown that they consist of trigonally connected $AsSe_3$ or $AsTe_3$ units, linked by short chains of selenium or tellurium atoms; the As—Se distance is 0·244 nm.[6] With increasing proportions of arsenic, the chains group together to form puckered layers, and an increasing fraction of the arsenic atoms become octahedrally co-ordinated. As was mentioned previously (Chapter 2, p. 26) arsenic lowers the polymerization temperature of sulphur and selenium, and Raman spectroscopy shows that at concentrations above 15 atom % arsenic, there are no longer any S_8 or Se_8 rings present in these polymers. It must be concluded that the chalcogenide glasses consist of short chains of chalcogen atoms, linked together by the multivalent atoms which act as branching and cross-linking points. Hence the parent network of the arsenic, antimony and phosphorus chalcogenides is a three-connected assembly of As, Sb or P atoms linked through chalcogen atoms (Fig. 3.1); but with increasing proportions of chalcogen, the chain length between cross-linking points increases and the cross-link density of the network decreases until it can be regarded as a partially cross-linked linear polymer. A three-component germanium—arsenic—tellurium glass has been shown[7] to contain units such as

```
   Te      Te
  /  \    /  \
       Ge
      /  \
    As    As
```

in which the Te—Te distance is 0·42 nm and the Te—Ge—Te angle is 110°.

Addition of even small amounts of any of the halogens except fluorine markedly lowers the melt viscosity of chalcogenide polymers, so that the halogen atoms evidently act as chain terminators. Hopkins[8] found evidence that when iodine is added to arsenic sulphide, some arsenic-sulphur bonds are severed and some arsenic-iodine bonds are formed. Since each arsenic-iodine

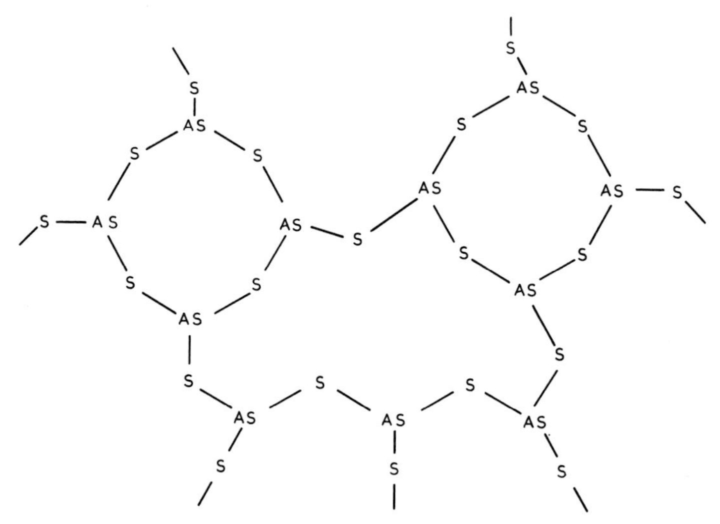

Fig. 3.1. A typical chalcogenide polymer network: arsenic sulphide glass.

bond must leave a singly bonded sulphur atom, which will immediately link with another such atom to form a disulphide link, the effect of iodine is not only to reduce the cross-link density but also to increase the average length of the relatively flexible chalcogen chain segments between branch-points. Both these changes will have the effect of reducing the melt viscosity.

Properties

The chalcogenide glasses are more like organic polymers than any other inorganic glasses. They have much lower softening points, higher refractive indices, and lower tensile strengths and elastic moduli than the oxide glasses. The properties of vitreous arsenic sulphide are typical of many simple chalcogenide polymers, and are listed in Table VII. The effects of variations in composition on some of these properties are discussed in the following sections.

Table VII
Properties of arsenic trisulphide glass

Glass transition temperature	200°C
Boiling point (760 mm)	735°C
Young's modulus	16 $GN m^{-2}$
Tensile strength (of fibre)	14–45 $MN m^{-2}$
Coefficient of linear expansion	$23{\cdot}7 \times 10^{-6}$ $°C^{-1}$
Specific resistivity	$8{\cdot}5 \times 10^{13}$ ohm cm
Refractive index (at 4 μm)	2·4
Transmission (2 mm thick)	70% from 2–10 μm

Glass Transition Temperatures

The glass transition temperatures of arsenic sulphide, arsenic selenide and germanium selenide glasses increase in approximately linear fashion with arsenic or germanium content.(Fig. 3.2). If every arsenic (or germanium) atom were a cross-linking point in the polymer network, the glass transition temperature would be related to the arsenic (or germanium) content by the Gibbs–di Marzio equation

$$T_x = T_0/(1-Kx)$$

where T_x, T_0 are the glass transition temperatures of the cross-linked polymer and the corresponding linear polymer respectively, and x is the cross-link density.[9] According to this relationship, however, the glass transition temperature should rise at an increasing rate as the cross-link density increases. A linear relation with arsenic content suggests that only some of the arsenic atoms are effective as crosslinking points, and that the fraction of the total arsenic atoms present that are effective cross-linking points diminishes with increasing arsenic content. Presumably the remainder are either simple branching sites or simply terminal groups. The fraction of cross-linking atoms can be calculated as follows, assuming the Gibbs–di Marzio equation to apply to these polymers: if y is the atom fraction of arsenic in the polymer and x the fraction of cross-linking sites, then

$$T_x = T_0/(1-Kx) = T_0 + Ay,$$

where A is the slope of the straight line obtained by plotting glass transition temperature, T_x, against arsenic content.

Combining these equations yields

$$x/y = A/K(T_0 + Ay),$$

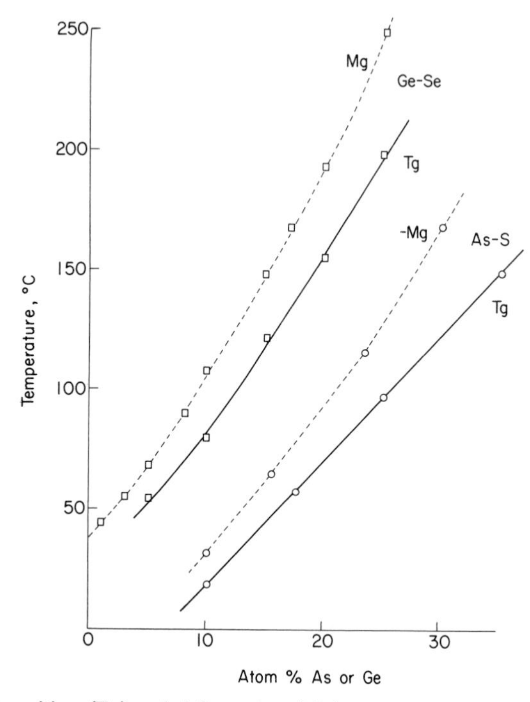

Fig. 3.2. Glass transition (Tg) and deformation (Mg) temperatures of arsenic sulphides and germanium selenides plotted against atom % cross-linking atoms. (From Holliday, L. (1975). "Ionic Polymers", Ch. 8. Applied Science Publishers, London. With permission.)

and by fitting experimental measurements of the glass transition temperatures of arsenic sulphides of known composition, the value of x/y for the arsenic-sulphur system is found to range from about 0·9 at $y = 0{\cdot}1$ to about 0·7 at $y = 0{\cdot}35$. That is, when the polymer contains only 19 atom % As, 90% of the arsenic atoms are effective as cross-linking sites; but in a polymer containing as much as 35 atom % As, only 70% of the arsenic atoms are behaving as true cross-linking points.

Viscoelastic Properties

A family of modulus-temperature curves for arsenic-sulphur and germanium-selenium glasses (Fig. 3.3) show that in the range 5–15 atom % arsenic and 3–8 atom % germanium the polymers have a distinct, though short, rubbery region; at 20 atom % As there is only an inflection in the modulus curve, and above 25 atom % As, no evidence for rubbery behaviour is apparent.[10] In this respect, the chalcogenide polymers differ markedly from the oxide (borate, silicate and

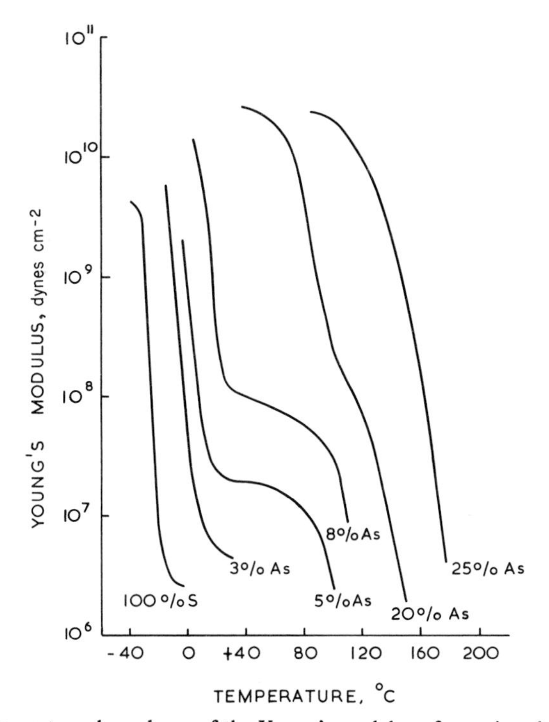

Fig. 3.3. Temperature dependence of the Young's modulus of arsenic sulphide glasses.

phosphate) glasses which do not appear to exhibit rubbery properties over any experimentally recorded range of temperature or composition.

Electrical Properties

The electrical conductivity of different chalcogenide polymers can vary over a wide range, as Table VIII shows.

The last three polymers in this list are intrinsic semiconductors, their conductivity increasing with temperature.

The conductivity of many chalcogenide glasses varies in an unusual way with applied voltage, undergoing reversible changes between a low conductivity and a high conductivity state. This phenomenon, known as "switching", can occur in two forms; threshold switching and memory switching (Fig. 3.4). In threshold switching, the glass maintains a high resistance until a certain threshold voltage gradient, typically 10^4 V cm^{-1}, reached; it then changes suddenly into a low resistance state with about a thousand-fold greater conductivity. It will remain in

Table VIII

Composition of polymer	Specific resistivity at 300 K, ohm cm
$Ge_{15}As_{15}Se_{70}$	5×10^{10}
$Si_{15}Sb_{33}S_{50}$	2×10^{9}
$Si_3Ge_2As_5Te_{10}$	1×10^{8}
$Ge\ As_4Te_5$	5×10^{5}
$Ge\ As_2Te_7$	$2{\cdot}8 \times 10^{4}$
$Tl_2As_2Te_3Se$	1×10^{3}

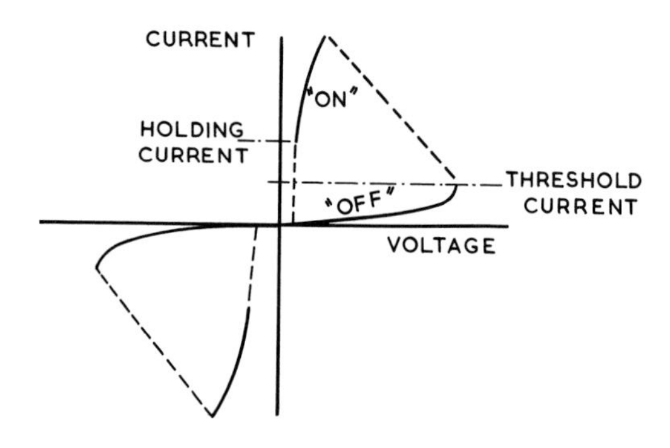

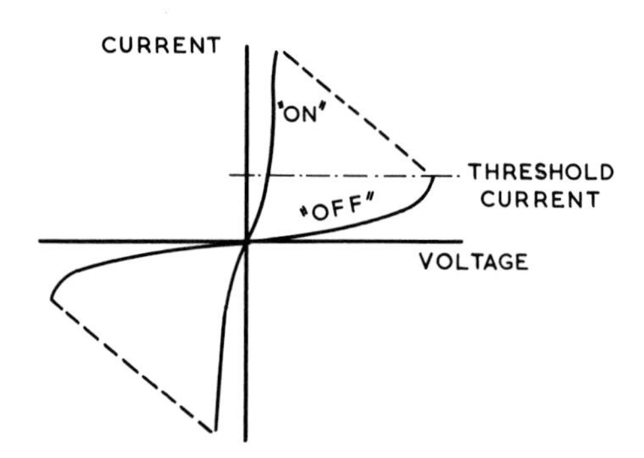

Fig. 3.4. "Threshold" and "memory" switching characteristics.

this condition so long as a certain minimum current (the "holding" current) is maintained; but if the current falls below this value the material returns, after a recovery period of the order of 10^{-6} s, to the high resistance state.

In memory switching, the glass exhibits a high resistance until a similar threshold voltage gradient is exceeded, when it changes suddenly into a low resistance state. It then remains in this condition even after the current is switched off; it can only be returned to the high resistance condition by passing a short current pulse that exceeds the original threshold voltage gradient.

When these effects were first observed there was much speculation and some controversy about the mechanism of the switching process. However, it now appears to have been established beyond reasonable doubt that the "leakage" current passing through the glass while in its high resistance state causes local decomposition, so that when the voltage gradient becomes high enough, a conducting path or filament is produced within the glass, connecting the electrodes. In the case of an As—Te—Ge glass, electron-probe microanalysis has shown that the conducting filament consists of crystalline As_2Te_3, the local germanium content being considerably reduced. Subsequent application of a high-current pulse heats the glass in the region of the conducting filament sufficiently to melt it, and re-vitrify the crystalline material. Measurements of the resistance at different temperatures show a similar effect; when the glass is heated to a certain threshold temperature, a crystalline phase begins to separate; its resistance suddenly drops to a low value which is maintained after cooling, and the relatively high conductivity of the crystalline region remains until the glass is subsequently heated above the melting point of the crystalline phase. The resistance returns to its original high value when the polymer is again vitrified by rapid cooling.

Some compositions exhibit both threshold and memory switching; in a glass of composition $Si_{12}Ge_{10}As_{30}Te_{48}$ the current at first rises linearly with applied voltage until, at the threshold voltage gradient, the current begins to rise very steeply. If the current—and consequently the heat dissipated—is limited by a suitable load resistor in the circuit and at the same time the applied voltage is reduced, the glass will return to its high resistance state and this cycle can be repeated many times. If, however, the current is allowed to rise to a higher value by reducing the load resistor, the glass will remain in the "on" condition after the current is switched off. It can subsequently be restored to its original state by passing a brief, high voltage pulse.

Applications

Chalcogenide glasses have been used as infra-red transparent windows for both military and civil devices, for example in the encapsulation of photo-sensitive transistors, and for construction of infra-red detectors. As amorphous semiconductors, it has been proposed to use them in ultrasonic delay lines, high energy-particle detectors and electron multipliers. Suggested applications of their switching properties have included memory devices for computers, electroluminescent displays, and so on. In fact, none of these electronic applications has been commercially successful, chiefly because of the poor reliability and variable characteristics of the switching devices that have been produced so far.

Ultraphosphate Glasses

Ultraphosphate glasses are amorphous polyphosphates in which the ratio of phosphate units to cation equivalents is greater than unity; they can be considered to be formally derived either from phosphorus pentoxide by interruption of a proportion of the P—O—P bonds by metal cations and hydroxyl groups, or from the linear polyphosphates $(MPO_3)_n$, by successive substitution of pairs of M^+ ions by an oxygen cross-link. The parent network is amorphous phosphorus pentoxide which consists of PO_4 units sharing three out of every four oxygen atoms to form a random, three-connected network (Fig. 3.5).

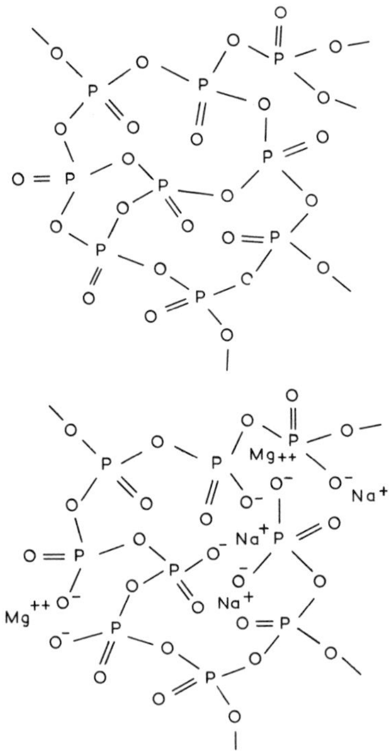

Fig. 3.5. Vitreous phosphoric oxide and a sodium-magnesium ultraphosphate glass.

Of all the oxide glasses, the ultraphosphates most closely resemble the glassy organic plastics in processability; they can have softening temperatures in the same range as some plastics and certain compositions can be processed by injection moulding, extrusion and vacuum-forming with the same machinery that is designed for processing plastics.

The techniques of polymer physics have been applied with notable success to the study of linear polyphosphates; by contrast, the highly cross-linked, insoluble polyphosphate glasses have received comparatively little detailed study. Recently it has been discovered[11] that some of these materials have novel and potentially valuable properties, especially in their unusual surface characteristics, which make them suitable for the production of non-misting spectacle lenses and anti-fouling surfaces for marine environments.

Precise structural analysis of these polymers is difficult because, being highly cross-linked, they cannot be dissolved without decomposition; their infra-red absorption spectra generally consist of a few very broad bands lacking in detail and the information available from X-ray diffraction is limited. Because of this, indirect methods have to be used, and Raman spectroscopy combined with thermal analysis and melt viscosity measurements have so far provided most of the useful information about these materials.

Preparation

Ultraphosphate glasses are made by similar methods to those employed for linear polyphosphates, except that additional phosphoric acid or other suitable precursor of phosphoric oxide, such as ammonium dihydrogen phosphate, is added to the reaction mixture to bring about the desired extent of branching and cross-linking in the polymer. For example, a cross-linked sodium ultraphosphate is prepared by heating a mixture of sodium dihydrogen phosphate and either phosphoric acid or ammonium dihydrogen phosphate in the molar ratio of $(1-x)/x$, where x is the cross-link density required, until the calculated amounts of water and ammonia have been eliminated. It is also possible to arrive at the same final product by heating together sodium carbonate and ammonium dihydrogen phosphate, when ammonia and carbon dioxide are eliminated first, followed by water as the temperature is raised:

(a) $NaH_2PO_4 + 2\ NH_4H_2PO \rightarrow [NaP_3O_8] + 2\ NH_3 + 4\ H_2O$;
(b) $Na_2CO_3 + 6\ NH_4H_2PO_4 \rightarrow 2[NaP_3O_8] + 6\ NH_3 + CO_2 + 9\ H_2O$.

Glasses of more complex composition can be prepared in a similar way by heating together mixtures of various metal oxides and carbonates with phosphoric acid or ammonium phosphate.[12,13] It has been found that the polycondensation reaction occurs in two stages; up to about 500°C the condensation proceeds only as far as the formation of a linear polyphosphate, the product at this stage being a water-soluble polymer containing acidic hydroxyl groups:

$$\begin{matrix} NaH_2PO_4 \\ + \\ 2NH_4H_2PO_4 \end{matrix} \xrightarrow[400\text{–}500^\circ C]{} \begin{matrix} & O & & O & & O & \\ & \| & & \| & & \| & \\ - & P & -O- & P & -O- & P & -O- \\ & | & & | & & | & \\ & O^- & & OH & & OH & \\ & Na^+ & & & & & \end{matrix} + 2NH_3 + 3H_2O$$

The subsequent elimination of a further molecule of water between each pair of hydroxyl groups, to form P—O—P cross-links between the chains, requires temperatures in excess of 600°C, and even at much higher temperatures complete dehydration of the polymer is difficult to achieve; indeed Namikawa and Munakata[14] concluded that the preparation of an anhydrous barium phosphate glass was impossible. The reason for this is that the reaction

$$2 >P(O)OH \rightarrow >P(O) \rightleftharpoons O—P(O)< + H_2O$$

is reversible, and because phosphorus pentoxide forms an azeotrope with water boiling at 850°C,[15] it is not possible to raise the temperature of the reaction mixture much above 800°C without altering the composition of the polymer by volatilization of phosphoric oxide. Nevertheless it is, in fact, possible to bring about almost complete dehydration of the polymer by adding dicyandiamide to the reaction mixture; this substance reacts rapidly and completely with hydroxyl groups attached to phosphorus forming ammonia and carbon dioxide, both of which are more readily and more completely eliminated from the reacting system than water and also avoid the complication of azeotrope formation.

By interrupting the polycondensation reaction at various intermediate stages before all the residual hydroxyl groups have been eliminated, polymers with different degrees of cross-linking, but with similar elementary compositions in terms of cation: phosphate ratio, can be isolated.[16] The cross-link density of an intermediate condensation product of this sort can be calculated from its elementary analysis, as follows:

If the mole fraction of phosphoric oxide in the polymer is y, the proportion of phosphate groups that are associated with cations is $(1 - y)/y$, so that the fraction capable of forming cross-links is $(2y - 1)/y$. If the cross-link density is x, then the fraction of phosphate groups carrying a residual hydroxyl group will be $[(2y - 1)/y] - x$, and this must be equal to the proportion of combined water left in the polymer, expressed as moles H_2O per mole P_2O_5. Now if M is the weight of anhydrous glass that contains one mole of P_2O_5, and z the fraction of phosphate groups carrying a hydroxyl group, the weight fraction of combined water is $18\ z/(M + 18\ z)$.

From these expressions it is evident that the cross-link density is given by

$$x = (2y - 1)/y - WM/18(1 - W),$$

where W is the weight fraction of combined water. The value of M is determined by the mole fraction of total phosphate in the reaction mixture (or by analysis of the polymer), and W can be measured either by a combustion analysis for hydrogen, or by measurement of the weight loss on ignition in the presence of an excess of a base that will react with, and so prevent any loss of, phosphoric oxide. For this purpose, litharge was found to be particularly suitable; the powdered glass is mixed with a considerable excess of litharge and heated at 700°C until there is no further weight change. The loss in weight is then equal to the combined water content, provided that the polymer is free from ammonium salts. If residual ammonium ion is present, a correction must be applied for the loss in weight due to the volatilization of ammonia.

Glass Transition Temperature

The glass transition temperatures of ultraphosphate glasses depend both upon their overall composition—that is, the kinds and proportions of different cations present—and also upon the actual cross-link density. The total molar fraction of cations in the polymer determines the maximum value of cross-link density and hence the highest Tg that can be reached with that composition; but if the polycondensation reaction is interrupted before the elimination of hydroxyl groups is complete, polymers of similar overall composition but with lower cross-link densities and hence lower glass transition temperatures will be obtained. A plot of maximum cross-link density against the mole fraction of P_2O_5 in the polymer (Fig. 3.6) shows that the range of actual cross-link densities that is accessible increases with P_2O_5 content; similarly the range of accessible glass transition temperatures must increase as the fraction of P_2O_5 in the polymer increases and its composition approaches closer to the parent network. Nevertheless the maximum glass transition temperatures of a series of ultraphosphate polymers containing increasing proportions of phosphoric oxide must nearly always decrease with P_2O_5 content, because the glass transition temperatures of most linear polyphosphates (see Table IV, p. 33) are higher than that of phosphoric oxide itself (273°C), and to a first approximation the maximum glass transition temperature of an ultraphosphate polymer is the weighted mean of the glass transition temperatures of the corresponding linear polyphosphate and phosphoric oxide. For example, a barium ultraphosphate polymer containing 60 mol % P_2O_5 can be considered to be a copolymer containing 40 mol of linear barium polyphosphate and 20 mol of fully cross-linked phosphoric oxide.

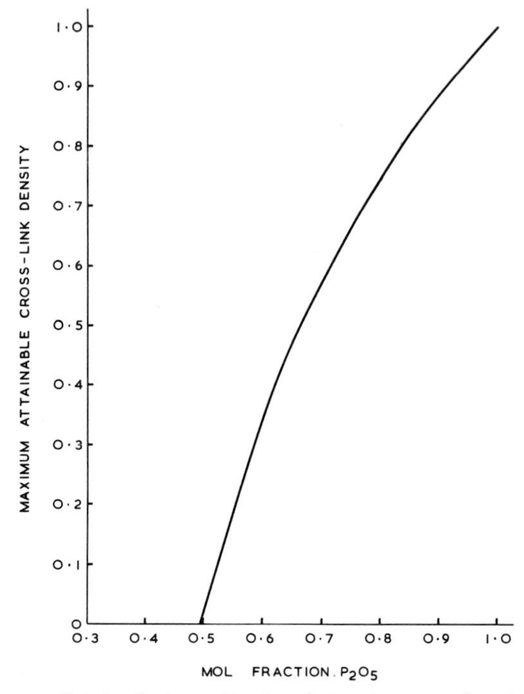

Fig. 3.6. Maximum cross-link density in an ultraphosphate polymer as a function of P_2O_5 content.

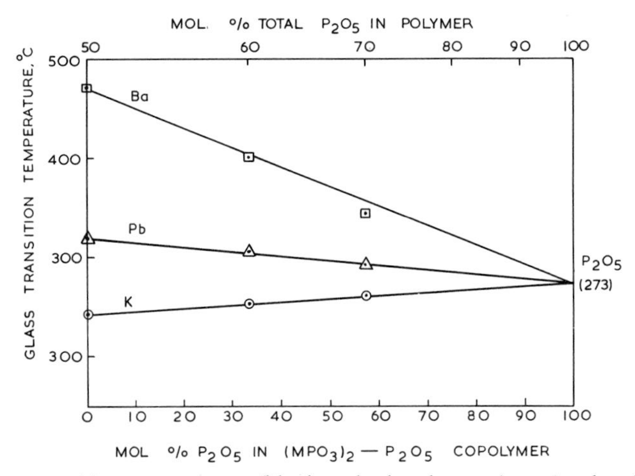

Fig. 3.7. Glass transition temperatures of barium, lead and potassium ultraphosphates as a function of P_2O_5 content.

On the assumption of a linear relation between the glass transition temperature and copolymer composition, this copolymer would have a glass transition temperature of 404°C; the actual value is 400°C. The glass transition temperatures of some typical metal ultraphosphates are shown in relation to phosphoric oxide content in Fig. 3.7; potassium and silver ultraphosphates are exceptional in that they have lower glass transition temperatures than phosphoric oxide. In systems with two or more cations, and especially in combinations of monovalent and divalent cations, the relationship between glass transition temperature and composition are more complex, and relatively few such systems have been studied in sufficient detail to be able to state anything more than general conclusions about the effects of particular cations. One that has is the system of ultraphosphate glasses containing lead and an alkali metal, which is interesting because it includes some polymers with glass transition temperatures in the range 140–250°C, which overlaps the range of softening temperatures of organic thermoplastics.[12] With a given ratio of lead to alkali metal, the glass transition temperature is lowest for potassium, and rises gradually with increasing lead content. If lead were behaving purely as a divalent cation in these networks, its effect would be to increase the proportion of ionic bonds between phosphate units in the network, and the softening temperature would be expected to rise much more steeply than is actually observed. Hassan and co-workers have suggested[17] from measurements of optical absorption and electron-spin resonance spectra that lead oxide can behave partly as a network-former in calcium-lead phosphate glasses, and this has been confirmed by the present author, using Raman spectroscopy to analyse the system of potassium-lead phosphate polymers.[18] The Raman spectra of all polyphosphates contain two prominent bands at 1155 cm^{-1} and 690^{-1}, which are attributable to vibrations of the PO_4 groups and the P—O—P bonds, respectively. In high molecular weight potassium polyphosphate, the intensities of these two bands are approximatel equal; but as increasing proportions of lead are substituted for potassium atoms in the polymer, the intensity of the 690 cm^{-1} band relative to the 1155 cm^{-1} band diminishes in an approximately linear fashion with lead content. This must mean that the ratio of P—O—P linkages to phosphorus atoms is decreasing; but the melt viscosity does not decrease, so that the molecular weight cannot be falling, and the most reasonable explanation is that some of the lead atoms are being incorporated into the polyphosphate chain, forming P—O—Pb linkages. There are, therefore, fewer ionic bonds between the chain segments and the glass transition temperature does not increase with lead content by nearly as much as would be expected. By contrast, the inclusion of even minor amounts of divalent metals such as magnesium, calcium, barium, cadmium or zinc raises the glass transition temperature of a polyphosphate considerably.[12]

Another system of ultraphosphate polymers that includes some glasses with unusually low softening points is the alkali-zinc phosphate system.[13] Although the glass transition temperature of linear zinc polyphosphate (530°C) is very much higher than that of lead polyphosphate (320°C), ultraphosphates containing up to 30 mol % zinc oxide together with a minor proportion of an alkali metal can have glass transition temperatures in the range 150–300°C. In this system the lowest glass transition temperatures were again found to occur when the alkali cation is potassium, and the glass transition temperature rises only gradually with increasing proportions of zinc, but additions of alkaline earth metal cations cause much larger increases in glass transition temperature. When there is more than 70 mol% of phosphoric oxide in the polymer, magnesium ions raise the softening point more than barium, but with phosphate contents less than 70 mol % the reverse is true (Table IX).

Table IX
Transformation temperatures of lead and zinc ultraphosphate polymers containing other cations

Other cation (R_2O or RO)	LEAD Polymers with 30 mol % PbO and 10 mol % R_2O or RO	ZINC Polymers with 10 mol % ZnO and 20 mol % R_2O or 10 mol % RO
Li	215	221
Na	220	188
K	190	183
Mg	293	236
Ca	241	
Ba	223	225

These apparently complex relationships between composition and glass transition temperature in the ultraphosphates can be rationalized in terms of ionic cross-linkages and oxygen density. Although the introduction of cations into a covalent oxide network such as phosphoric oxide reduces the cross-link density in terms of covalent bonds, if this were the only effect, then all cations would produce the same change in transformation temperature. This is not so, and the reason is that all cations, even those of the alkali metals, form co-ordinate linkages to several oxygen atoms in the network, the number increasing with ionic radius; and for cations of the same valency, the strength of these bonds decreases with the size of the cation (Table X). Although these ionic forces are weaker than the covalent links they replace, they tend to counteract the effects of the reduction in cross-linking. The smaller and more highly charged cations cause a tightening-up of the network and hence a reduction in free volume that is reflected in a higher glass transition temperature; while the expansion of the

Table X
Co-ordination number and oxygen bond strength of some cations on oxide networks

Cation	Co-ordination number	Single bond strength to oxygen, $kJ\,mol^{-1}$
Li	4	151
Na	6	84
K	9	54·5
Rb	10	50
Cs	12	42
Mg	6	155
Ca	8	134
Sr	8	134
Ba	8	138

network that is required to accommodate the larger and less highly charged ions increases the free volume and consequently lowers the glass transition temperature. One way of characterizing the tightness of packing, or packing density, of an oxide network is in terms of its oxygen density—that is the weight of oxygen contained in unit volume of polymer. The higher the oxygen density of the network, the more closely packed are the oxygen atoms and the greater is the internal energy required to obtain chain mobility; thus for a given cross-link density, the glass transition temperature increases with increasing oxygen density. Secondly, for a given oxygen density, the glass transition temperature is higher with divalent cations than with monovalent cations. Thirdly, for a given composition with specified cations and a particular ratio of cations to total phosphate the cross-link density can be varied by interrupting the polycondensation reaction at different stages, so as to leave unreacted hydroxyl groups in the polymer, and the glass transition temperature then varies with cross-link density according to the Gibbs–di Marzio relationship(9)

$$T_x = T_0/(1 - Kx),$$

where T_x, T_0 are the glass transition temperatures of the cross-linked polymer and the corresponding linear polymer respectively, K is a constant, and x is the cross-link density.

These relationships are illustrated in Figs 3.8 and 3.9, which show some of the experimental results obtained by the author for a number of multicomponent ultraphosphate glasses.(12, 13) In Fig. 3.8, the measured glass transition temperatures of three different ultraphosphate compositions are plotted against cross-link density at different stages during the polycondensation. The curves drawn in this diagram are calculated from the Gibbs–di Marzio equation, using the

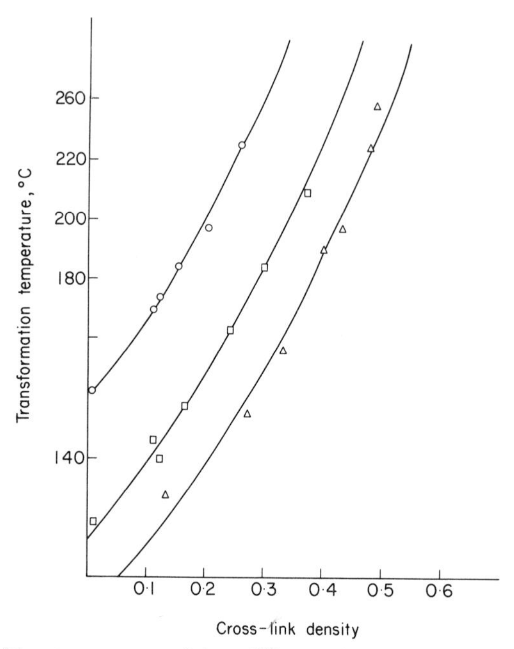

Fig. 3.8. Glass transition temperatures of three different ultraphosphate glasses plotted against cross-link density; experimental results superimposed on curves calculated by the Gibbs–di Marzio equation: (O) glass 1, 60 mol % P_2O_5; (□) glass 2, 65 mol % P_2O_5; (Δ) glass 3, 68 mol % P_2O_5. (From Ray, N. H. (1974) *J. Non-crystalline Solids* **15**, 429, with permission.)

observed values of T_0 in each case and a common value of 0·6 for K; they all fit the experimental points remarkably well. The three compositions differ in phosphoric oxide content, and hence in maximum attainable cross-link density; they also differ in oxygen density as a result of differing proportions of cations. The glass with the highest phosphoric oxide content, glass 3, contains potassium as the predominant cation, while the other two contain increasing proportions of lithium and magnesium ions respectively, so that the oxygen densities increase from glass 3 to glass 1. For this reason the glass transition temperatures at any given cross-link density also increase from glass 3 to glass 1 but, because the maximum attainable cross-link density is highest for glass 3, the glass transition temperatures of the most highly condensed samples of these polymers increase from glass 1 to glass 3.

In Fig. 3.9, the measured glass transition temperatures of a number of different ultraphosphate polymers are plotted against oxygen density. These polymers fall into three groups: lead ultraphosphates containing 30 mol % PbO and 10 mol % of another cation; zinc ultraphosphates containing 20 mol % ZnO and from 9–15 mol % of other cations; and simple two-component ultraphosphates containing between 55 and 60 mol % P_2O_5. The maximum cross-link densities of these three groups are therefore 0·33, 0·46–0·59, and 0·18–0·33 respectively. Although the individual cross-link densities of all these polymers vary according to reaction conditions over a range that is determined principally by their composition, the general trend of increasing glass transition temperature with increasing oxygen density is apparent.

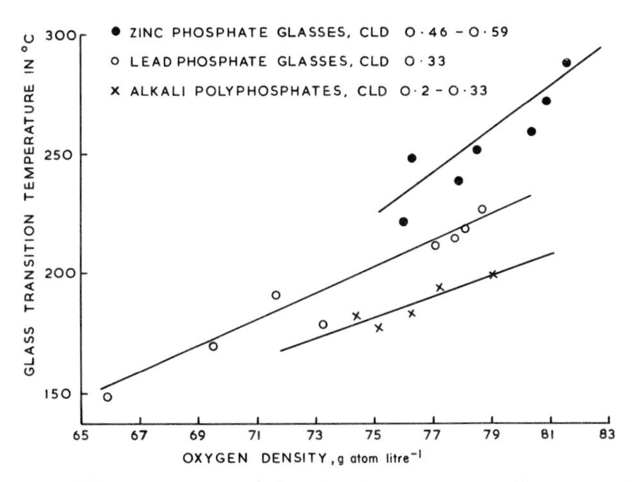

Fig. 3.9. Glass transition temperatures of ultraphosphate glass with different cross-link densities plotted against oxygen density.

Durability

The P—O—P linkages in a simple polyphosphate are thermodynamically unstable towards hydrolysis, the equilbrium

$$>P(O)—O—P(O)< + H_2O \rightleftharpoons 2 >P(O)OH$$

lying well over to the right, as the following enthalpies of reaction in Table XI indicate.

Table XI

Reaction	ΔH, K cal mol^{-1} P—O—P bonds severed
$(HPO_3)_{2n} + n\ H_2O \rightarrow n\ H_4P_2O_7$	−3·9
$H_4P_2O_7 + H_2O \rightarrow H_3PO_4$	−6·1

The rate at which hydrolysis occurs depends on a number of factors, among which the rate of diffusion of water into the polymer and the solubility and rate of removal of the hydrolysis products are probably the most important. All the simple alkali metal ultraphosphates are rapidly attacked and dissolved by water at room temperature, the rate of attack increasing in the order $\mathrm{Li}<\mathrm{Na}<\mathrm{K}$, which is also the order of decreasing oxygen density. Polyphosphates containing alkaline earth metal cations are much more resistant to hydrolysis, magnesium ultraphosphates being more durable than barium ultraphosphates for the same metal : phosphorus ratio. The durability of ultraphosphate glasses also increases with cross-link density,[12] and these observations suggest that the more tightly bound is the network, the more resistant it should be to hydrolytic attack. This is not always the case, however, for zinc ultraphosphate glasses with oxygen densities in the range 80–84 g atom litre^{-1} are less durable than lead ultraphosphates with oxygen densities around 70 g atoms litre.[13] Evidently lead, which can behave as an additional network-former in phosphate glasses (see preceding section), has a greater effect in reducing the rate of hydrolysis of the network than a simple cation. The durability of ultraphosphate polymers is also strongly pH dependent; Kamazawa and co-workers[19] found that the rate of attack by water on the alkaline earth ultraphosphates $RO\cdot(P_2O_5)_x$ was most rapid in neutral solution for $x = 1$, that is at the metaphosphate composition. In acidic solution the rate of attack decreases as x increases, while in alkaline media, the reverse is true.

By taking account of these general principles it is possible to select combinations of cations that will improve the durability of ultraphosphate glasses by as much as two orders of magnitude. Since the same parameters that tend to reduce the rate of attack by water also tend to raise the glass transition temperature, there is inevitably an inverse correlation between rate of dissolution in water and glass transition temperature.[13] Plotting either the reciprocal or the logarithm of the rate of solution against glass transition temperature for a number of different kinds of ultraphosphate glasses yields a series of points scattered about one or two straight lines that are characteristic of each type of ultraphosphate composition (Fig. 3.10). It follows that for applications in which a low softening point is advantageous—for example in injection moulding and extrusion—a valid comparison of different ultraphosphate compositions on grounds of durability must take the glass transition temperature into account; an appropriate "figure-of-merit" might be the reciprocal of the product of the glass transition temperature (in °C) and the rate of solution (in mm day^{-1}). Table XII gives some typical values for a number of ultraphosphate glasses. It can be seen that lead ultraphosphate glasses are more durable than zinc ultraphosphates of the same glass transition temperature; but no composition of this type that has a sufficiently low softening point to be injection-moulded or extruded in plastics processing equipment—that is with a glass transition temperature below 200°C—is likely

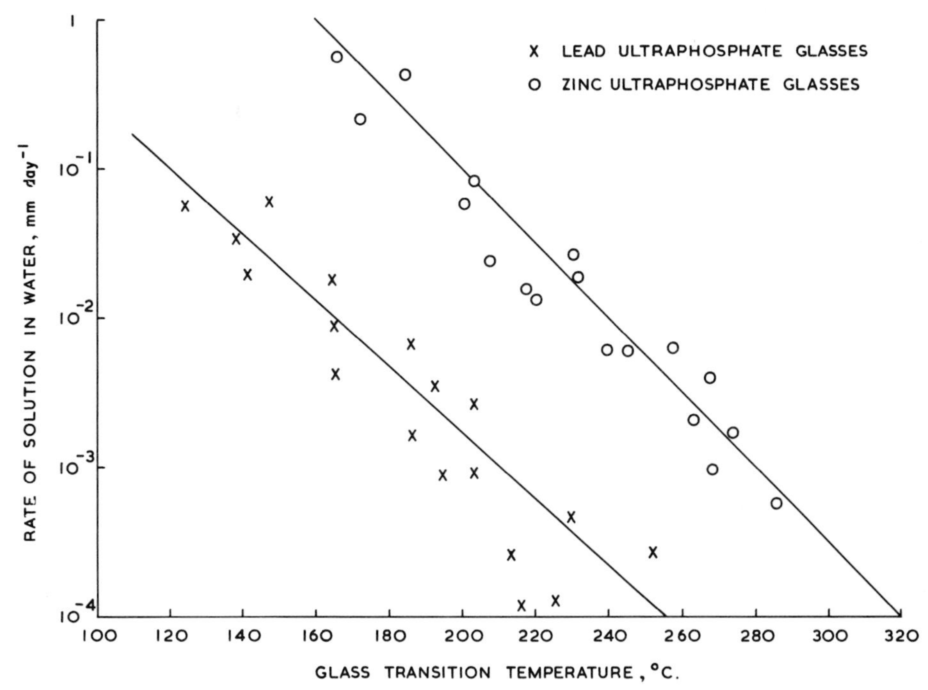

Fig. 3.10. Durability (expressed as rate of solution in water) of ultraphosphate glasses as a function of glass transition temperature.

Table XII
Comparative durabilities of some low-softening point oxide glasses

Glass type	Tg, °C	Rate of solution in water mm day^{-1} at 20°C	100°C	Figure of merit
Lead utraphosphate	205	$1{\cdot}9 \times 10^{-4}$	1·7	25·7
Lead metaphosphate	200	$3{\cdot}3 \times 10^{-4}$		15·2
Zinc ultraphosphate	225	$2{\cdot}6 \times 10^{-3}$		1.7
Alkaline earth borophosphate	330	$1{\cdot}1 \times 10^{-5}$	0·07	275
Lead borate solder glasses	394	$2{\cdot}5 \times 10^{-5}$	$5{\cdot}2 \times 10^{-4}$	101
	350	$9{\cdot}1 \times 10^{-5}$	$1{\cdot}2 \times 10^{-3}$	31
Silicate glasses:				
S95	575	$1{\cdot}9 \times 10^{-5}$	9×10^{-4}	92
Optical crown	600	7×10^{-6}	$6{\cdot}5 \times 10^{-4}$	238

[a] "Figure of merit" is inversely proportional to the product of the transformation temperature in °C and the rate of solution in mm day^{-1} at 20°C.

to be durable enough to be used for external components like car headlamp glasses. The improvements in durability that can be obtained by different permutations of cations are evidently limited; much greater improvements can however be achieved by introducing additional glass-forming (polymeric network) oxides into the polymer (see Chapter 4).

Melt Viscosity

The melt viscosities of the ultraphosphate polymers are practically independent of shear rate; in this respect, the cross-linked polyphosphates differ markedly from the high-molecular weight linear polyphosphates. This indicates that the average length of the mobile segments in a molten ultraphosphate is below the critical value at which non-Newtonian flow behaviour commences. For a great many polymers (including the few inorganic polymers for which such data are available) the average chain length at which the melt viscosity becomes shear-dependent is approximately 14 M/d, where M is the molecular weight of the repeat unit and d is the density of the melt. If this relationship holds true for polyphosphates, it means that the average length of a mobile segment of the network that is concerned in viscous flow is less than 150 repeat units, i.e. PO_4 units. Evidently in the molten state a high proportion of the cross-links in the network must be undergoing bond interchange, thus setting up a dynamic equilibrium concentration of mobile links that permits flow to occur by successive small movements of comparatively short segments. The temperature dependence of viscosity of these polymers is quite closely described by an equation of the form

$$\eta = A \exp (E/RT).$$

For linear polyphosphates such as sodium polyphosphate, Van Wazer[15] found that the activation energy E increased as the Na_2O/P_2O_5 ratio decreased towards the metaphosphate composition—that is the activation energy increased with molecular weight. Beyond the metaphosphate composition there have been comparatively few detailed measurements; however, the present author found that in a series of multicomponent ultraphosphate glasses containing from 60–70 mol % P_2O_5, the activation energy of viscous flow was practically independent of cross-link density[13] and very close to the value for molten phosphoric oxide, the parent network polymer (174 kJ mol^{-1}). However, the actual viscosity of any particular composition at a constant number of degrees above the glass transition temperature was found to increase markedly with cross-link density; unless, therefore, there is an abrupt change in activation energy near to the glass transition, the glass transition temperature cannot be an isoviscous temperature for this family of polymers.

Modulus

The Young's modulus of most three-connective oxide glasses (borates, arsenates, phosphates, etc.) lies in the range 20–55 GN m^2. Typical cross-linked ultraphosphates—for example a glass containing 60 P_2O_5, 25 PbO, 10 K_2O, 5 BaO—have moduli at the upper end of this range (5 1 GN m^{-2} for the composition quoted). Starting from a linear polymer containing unreacted hydroxyl groups, the modulus at first hardly changes as the condensation is continued, until a cross-link density of about 0·25 is reached, after whch the modulus increases rapidly with cross-link density (Fig. 3.11). As with other oxide

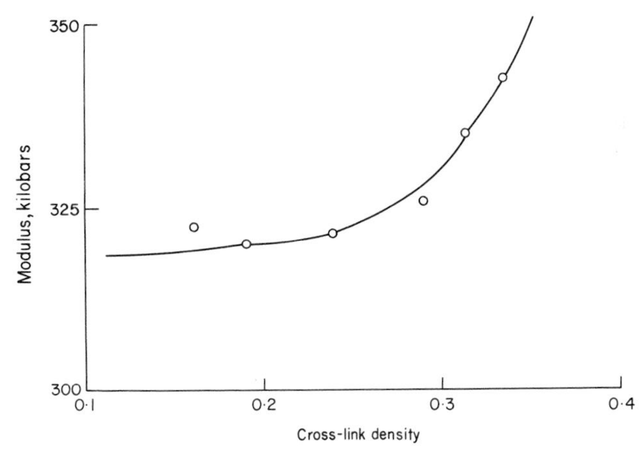

Fig. 3.11. Young's modulus of an ultraphosphate glass as a function of cross-link density. (From Ray, N. H. (1974). *J. Non-cryst. Solids,* **15,** 433. With permission.)

network polymers, the elastic modulus of fully cross-linked ultraphosphate glasses is approximately proportional to their oxygen density (Fig. 3.12). It follows that bulky cations of low charge that form comparatively weak bonds to oxygen reduce the modulus, while small cations of high charge that form strong bonds to oxygen increase the modulus. Thus the highest moduli are to be found among cross-linked phosphate polymers which contain ions such as lithium and magnesium.

Surface Properties

The surface of an ultraphosphate glass in a normal atmosphere is covered with a very thin layer of partially hydrolysed glass and adsorbed water at an equilibrium concentration that depends upon the humidity. This layer makes the glass

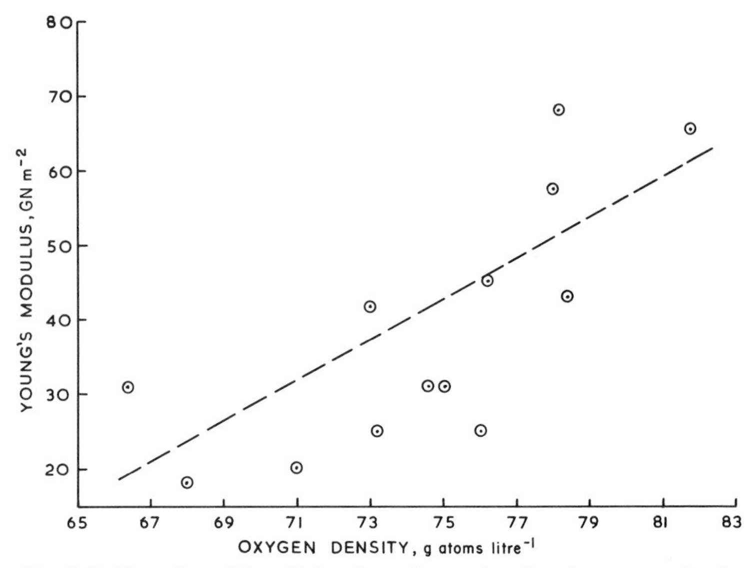

Fig. 3.12. Young's modulus of phosphate glasses plotted against oxygen density.

non-misting, that is, when a cold glass surface is brought into a warmer humid atmosphere, the water vapour that condenses on the glass forms a uniform thin film rather than discrete droplets, so that vision through the glass in unimpaired. This phenomenon is discussed more fully in the section on borophosphate glasses (Chapter 4), which have the same property but are considerably more durable, and are therefore more suitable for practical applications of this kind.

Boron Nitride

Boron nitride, which has the empirical formula BN, occurs in two forms, of which the commonest is a three-connective layer polymer resembling graphite, and the other is a crystalline, four-connected network polymer like diamond. It was first discovered in 1842 by Balmain[20] but the cubic, diamond-like, variety was not obtained until 1957.[21] The commoner three-connective polymer can be obtained by thermal decomposition of a wide variety of compounds and mixtures of reagents; for example by heating boric acid with urea in a nitrogen atmosphere at 600°C, followed by treatment of the white, cinder-like product at 950°C in a current of ammonia.[22] The final product is an amorphous powder which can be sintered above 1000°C to give dense, intact articles composed of aggregates of minute hexagonal crystals which average a few tens of nanometres in size. Another form of crystalline boron nitride which has rhombohedral symmetry has been obtained by reaction of potassium cyanide with boric oxide or borax at 1100°C in a graphite container.

The structure of hexagonal (three-connective) boron nitride resembles that of graphite in that it consists of infinite sheets made up of condensed six-membered rings of alternating boron and nitrogen atoms, stacked in layers (Fig. 3.13), but there is one important difference: in boron nitride the sheets are stacked so that the atoms in each successive layer are directly superimposed, while in graphite the atoms in each layer lie opposite to the centres of the six-membered rings in the next layer.[23]

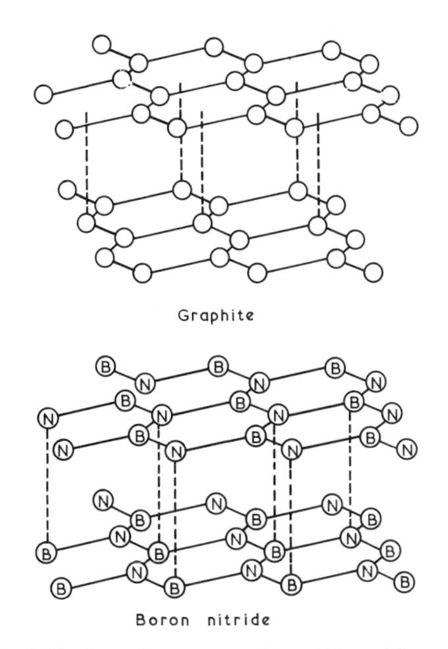

Fig. 3.13. Crystal structures of graphite and boron nitride.

The crystallographic cell constants of boron nitride are very similar to those of graphite, the B—N distance being 0·145 nm (C—C in graphite is 0·142 nm) and the interlayer separation is 0·33 nm (0·005 nm less than that in graphite).

Another important difference between boron nitride and graphite lies in their electronic structure. Unlike graphite, boron nitride is white and is an electrical insulator. According to Korshak and Mosgova[24] the electrons in boron nitride are highly localized as a result of the way in which the layers are stacked, so that a nitrogen atom in one layer is always adjacent to a boron atom in the next one. Consequently the electrons are bound between the layers, and the polymer is non-conducting.

Like graphite, boron nitride is soft and easily machined. Articles fabricated from this polymer, either by sintering or by machining, can be used in air up to about 800°C or in inert atmospheres up to 1600°C. Under nitrogen pressure the crystalline melting point is about 3000°C, but at atmospheric pressure the material sublimes below this temperature.

The transformation of the three-connected hexagonal boron nitride into the four-connected cubic form takes place on heating at 1500–2000°C and 50–90 kbar in the presence of catalysts. The catalysts that are customarily employed in the conversion of graphite into diamond, such as iron, nickel and manganese, have no effect on the boron nitride transformation. Instead, substances such as the alkali metals, their nitrides, some alkaline earth metals, lead, tin and antimony have been used. The transformation can also be effected without catalysts, but the lowest pressure which can then be employed is 110–120 kbar.

The cubic, four-connective form of boron nitride is isomorphous with diamond and for all practical purposes indistinguishable from diamond in hardness; it has been employed as a substitute for diamond both in cutting tools and in jewellery.

References

1. Schultz–Sellack, C. (1870). *Ann. Phys. Chem.* **139**, 182.
2. Frerichs, R. (1953). *J. Optical Soc. Am.* **43**, 1153.
3. Fraser, W. A. (1953). *J. Optical Soc. Am.* **43**, 823.
4. Pearson, A. D., Northover, W. R., Dewald, J. F., and Peck, W. F. Jr. (1962). "Advances in Glass Technology," Pt. 1, pp. 357–365. Plenum Press, New York.
5. Vaipolin, A. A. and Porai-Koshits, E. A. (1961). *Soviet Phys. Solid State,* **2**, 1500.
6. Leadbetter, A. J. and Apling, A. J. (1974). *J. Non-cryst. Solids,* **15**, 250.
7. Fitzpatrick, J. R. and Maghrabi, C. (1971). *Phys. Chem. Glasses,* **12**, 105.
8. Hopkins, T. E., Pasternak, R. A., Gould, E. S. and Herndon, J. R. (1962). *J. Phys. Chem.* **66**, 733.
9. Di Marzio, E. A., and Gibbs, J. H. (1972). *J. Polymer Sci.* **7**, 47.
10. Kurkijan, C. R., Krause, J. T. and Sigety, E. A. (1971). *IXth International Congress on Glass,* 1.4.
11. Ray, N. H., Lewis, C. J. and Robinson, W. D. (1973). British patent 1, 404, 914.
12. Ray, N. H., Lewis, C. J., Laycock, J. N. C. and Robinson, W. D. (1973). *Glass Technology,* **14**, 50.
13. Ray, N. H., Laycock, J. N. C. and Robinson, W. D. (1973). *Glass Technology,* **14**, 55.
14. Namikawa, H. and Munakata, M. (1965). *J. Ceram. Assoc. Japan,* **73**, 86.
15. Van Wazer, J. R. (1968). "Phosphorus and its Compounds". Interscience, New York.
16. Ray, N. H. and Lewis, C. J. (1972). *J. Materials Sci.* **7**, 47.
17. Hassan, F., Farid, S. and El Dabh, M. (1974). *Xth International Congress on Glass,* **I**, 5–16.
18. Ray, N. H. (1975). *Glass Technology,* **16**, 107.

19. Kamazawa, T., Ikeda, M. and Kawazoe, H. (1969). *J. Ceram. Assoc. Japan,* **77,** 163.
20. Balmain, W. (1842), *Phil. Mag.* **21**, 170.
21. Wentorf, R. H. (1957), *J. Chem. Phys.* **26**, 956.
22. O'Connor, T. E. (1962). *J. Amer. Chem. Soc.* **84**, 1753.
23. Bundy, F. P. and Wentorf, R. H. (1963). *J. Chem. Phys.* **38**, 1144.
24. Korshak, V. V. and Mosgova, K. K. (1959). *Uspeki Khimii,* **28**, 783.

4

Networks of Mixed Three- and Four-connectivity

Borate Glasses

Boric oxide itself is a three-connective network polymer, and polymeric glasses are readily formed by reaction of boric oxide with a wide variety of metal oxides or carbonates, including all the alkali metal and alkaline earth oxides, as well as zinc, cadmium, lead, bismuth, thallium and lanthanum oxides. When metal oxides are added to boric oxide, a proportion of the boron atoms become four-connected and acquire a negative charge that is balanced by the cation. For this reason the polymeric borate glasses must be classified as having networks of mixed three- and four-connectivity. Because of the comparatively low softening points of borate glasses they are used mainly as solder glasses for joining other glasses, such as parts of television tubes, and as sealing glasses for sealing metal and ceramic components into other glasses.

Formation

Boric oxide glass itself is obtained by dehydration of boric acid. This is a polycondensation reaction that results in a gradual increase in cross-link density and consequently a continually increasing viscosity. The last traces of water are very difficult to remove, even at temperatures as high as 1200°C. Metal polyborates are formed by melting together mixtures of metal oxides (or carbonates) and boric acid or boric oxide. The composition ranges over which stable, homogeneous polymeric glasses are formed have been determined for a number of different metals, for example by Imaoka,[1] and it is significant that zinc, cadmium, lead and bismuth oxides are capable of forming stable glasses that contain considerably less than 50 mol% of boric oxide. Stanworth[2] suggested that these oxides must be incorporated into the network, and this has been confirmed by nuclear magnetic resonance measurements[3] which show that with increasing lead content, the Pb^{207} chemical shift in a lead borate glass decreases from 98 gauss, corresponding to Pb^{2+} ions, towards 70 gauss which corresponds to a covalently bound lead atom as in litharge. Such polymers should therefore be regarded as copolymers of the metal borate and the metal oxide even though one of the components is not normally polymeric, as in the case of the zinc borates.

Properties

The glass transition temperaure of boric oxide glass has been reported by various authors to be from about 220° to 260°C. Since it is very difficult to obtain perfectly anhydrous boric oxide, and traces of water inevitably result in low values for the glass transition, the higher values are more likely to be correct; a very careful determination by Parks and Spaght[4] gave 247°C, which is probably near the right value.

As increasing proportions of alkalies are added to boric oxide, the glass transition temperature of the polymer rises steeply at first up to about 20 mol% alkali, and then levels out (Fig. 4.1). Many other properties of alkali borate glasses (such as density, melt viscosity and thermal expansion coefficient) also show apparently anomalous changes in behaviour with increasing alkali content at about the same molar proportion of alkali; for example, the thermal expansion coefficient passes through a minimum at around 15–17 mol % alkali.

Because this kind of behaviour is not commonly met with in systems based on other glass-forming oxides such as the silicates and phosphates, it has been referred to as the "borate anomaly" and was, at one time, the subject of much speculation. It must now be recognized, however, that comparatively sudden changes in the relationships between certain physical properties of glasses and their compositions in particular ranges are neither peculiar to the borate system, nor are they necessarily anomalous. The simple binary alkali silicate glasses also show sudden changes in melt viscosity and expansion coefficient in the region of 10 mol % alkali (see Chapter 5), and at least a basic understanding of these phenomena has been reached in most cases, if not a complete explanation.

Nuclear magnetic resonance measurements,[5] infra-red[6] and Raman spectra[7] have shown that the effect of increasing the alkali content of a borate glass is to introduce an equivalent proportion of tetravalent boron atoms, so that the structure in the neighbourhood of the cations changes from a three-connected network to a four-connected network (Fig. 4.2), the resulting negative charge on the tetravalent boron atoms being balanced by the associated cations. However, it has been found that crystalline compounds such as lithium metaborate contain only trigonally connected BO_3 units. The same change in boron co-ordination number is also evident in silver borate glasses[8] and lead borate glasses,[9] and

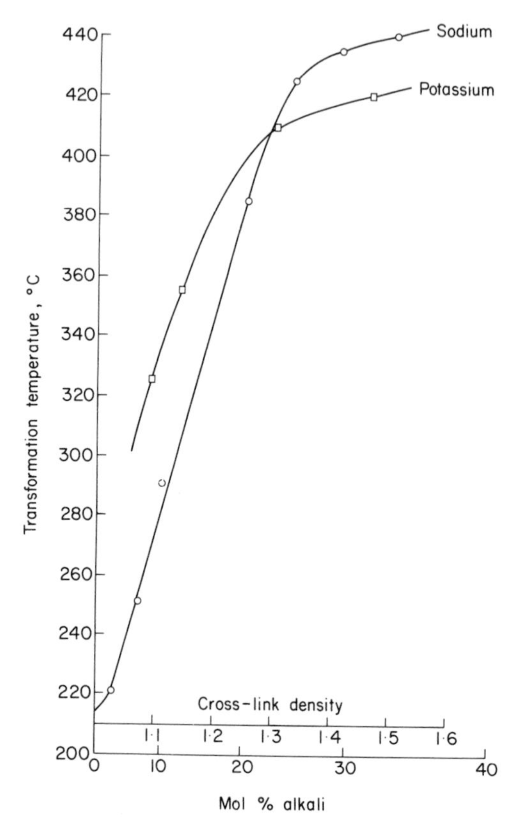

Fig. 4.1. Glass transition temperatures of sodium and potassium borate glasses as a function of cross-link density. (From Holliday, L. (1975). "Ionic Polymers", Ch. 8. Applied Science Publishers, London. With permission.)

the nuclear magnetic resonance results for all these systems indicate that up to about 30 mol % cations, the fraction of four-co-ordinate boron atoms is given by the expression

$$N_4 = \frac{\alpha}{2}\frac{x}{(1-x)},$$

where x is the mole fraction of alkali, silver or lead oxide, and α is the "conversion ratio", defined as the number of BO_3 units converted to BO_4 units per atom of oxygen added as metal oxide (Fig. 4.4, p. 78). For the alkali metal cations and silver, α is very close to 2, while in the case of lead borate glasses, the value of α decreases with increasing lead content.

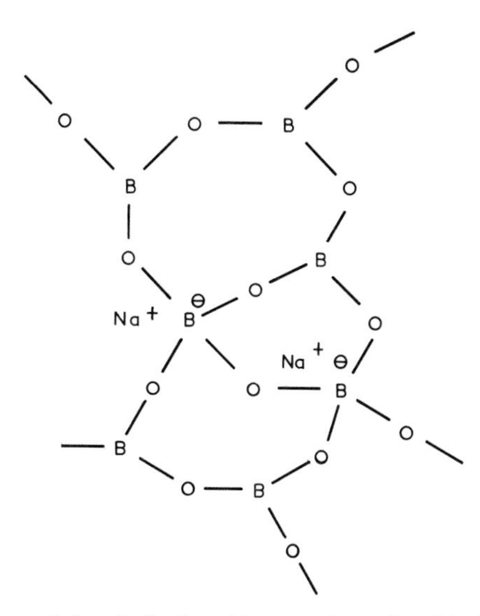

Fig. 4.2. Change in connectivity of a boric oxide network produced by introduction of cations.

In these networks the effective cross-link density (CLD) is most conveniently defined as the fraction of boron atoms linked to more than two other BO_3 or BO_4 units, and is given by the equation

$$\mathrm{CLD} = \frac{1 - x(1 - \alpha/2)}{1 - x}$$

which, for the alkali cations up to $x = 0{\cdot}3$, simplifies (since $\alpha = 2$) to

$$\mathrm{CLD} = \frac{1}{1 - x},$$

so that the cross-link density is numerically equal to the reciprocal of the mole fraction of boric oxide. Up to about 20 mol % alkali, the glass transition temperatures of sodium and potassium borate glasses increase rapidly with cross-link density (Fig. 4.1, p. 74). The available experimental data are probably not sufficiently precise to discriminate between a linear relationship (as suggested in the figure) and the curve predicted by the Gibbs–di Marzio equation. At higher alkali contents the glass transition temperature rises much more slowly, although the cross-link density continues to increase up to 30 mol % alkali in direct proportion to alkali content, as indicated by the fraction of four-co-ordinate boron atoms. This must indicate a change of structure, and it is significant that, up to about 20 mol % alkali, potassium borate glasses have higher glass transition temperatures than sodium borate glasses of the same molar ratio, whereas above 23 mol % alkali, this order is reversed. Furthermore, the change in slope of the curve relating glass transition temperature to alkali content occurs later—that is at a slightly higher alkali content—with sodium borates than with potassium borates. Since both cations produce the same change in cross-link density over the range of compositions concerned, this difference in behaviour is most likely to be due to differences in network packing density. Calculations show that below 20 mol % alkali, potassium borate glasses have the higher oxygen density, while at higher alkali contents, sodium borate glasses are more tightly packed. These differences indicate that there are probably a limited number of sites where alkali cations can be accommodated without distortion of the network. When all such sites are occupied, any further increase in alkali content brings about a structural change that results in a more open network, and the resulting increase in free volume tends to counteract the effects of the further increase in cross-link density on the glass transition temperature. If this is the case, it is to be expected that the larger cation, potassium, would cause expansion of the network at a lower concentration than the smaller sodium ion, and that the glass transition temperature would rise more due to increasing cross-linking with the smaller ion before the effects of reduced packing density became appreciable. This is precisely what is observed.

Expansion Coefficient

In most typical oxide glasses, such as the silicates and the phosphates, the introduction of alkali cations (except lithium) produces an increase in thermal expansion coefficient that can be attributed to the more open network and the reduction in covalent bonding. Lithium ions produce an increased network density in phosphate glasses and accordingly reduce the coefficient of expansion. When alkali cations are introduced into boric oxide glass, however, an unusual change in expansion coefficient is observed. Up to about 15 mol % alkali the expansion coefficient decreases for all alkali cations. Round about 15–17 mol % alkali the expansion coefficient reaches a minimum value and then begins to increase again, rising as far as the limit of the glass-forming range of compositions (Fig. 4.3). Many different explanations of these changes in expansion coefficient have been proposed. Before Bray's measurements by nuclear magnetic resonance techniques were available[5] the occurrence of a minimum in the expansion coefficient was thought to be associated with the change in co-ordination number of boron atoms from 3 to 4. It was believed that the addition of alkali oxides to boric oxide glass initially produced only four co-ordinate boron atoms, causing a tightening of the network and a decrease in expansion coefficient; then at about 15 mol% alkali, a limit was reached to the proportion of four-co-ordinate boron atoms that could be accommodated, and further additions of alkali produced non-bonding oxygen atoms associated with the three-co-ordinate boron atoms. This resulted in a loosening of the network and a consequent increase in expansion coefficient. Abe[10] and Huggins[11] suggested that the limit to the fraction of four-co-ordinate boron atoms in the glass corresponded to a network composed of BO_4 tetrahedra each surrounded by four

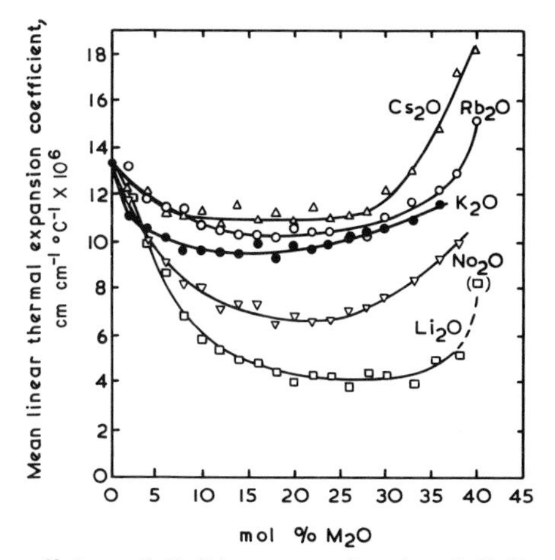

Fig. 4.3. Expansion coefficients of alkali borates as a function of alkali content. (From Uhlmann, D. R. and Shaw, R. R. (1969). *J. Non-cryst. Solids,* **1**, 347. With permission.)

BO_3 triangles, on the assumption that two BO_4 groups could not share a common oxygen atom because of their negative charge. This condition would be reached at the composition $Na_2O \cdot 5B_2O_3$, corresponding to 16·7 mol % alkali. Many other models have been proposed to account for the observed changes in expansion coefficient; for example Block and Piermarini[12] proposed a structure consisting six-, eight- and twelve-membered boroxole rings, by analogy with the structure of anhydrous crystalline borates. Krogh-Moe[13] presented a model involving distorted tetrahedra with what amounts to a variable co-ordination number over the range 0–33 mol % alkali. None of these theories can satisfactorily account for all the experimental data. In particular, Bray's nuclear magnetic resonance measurements, [5] supported by Raman spectroscopy and X-ray diffraction studies on crystalline borates[14] demonstrate conclusively that the fraction of four-co-ordinate boron atoms continues to increase in approximately linear fashion with alkali content at least up to 30 mol % alkali, so that the minimum in the expansion coefficient cannot be attributed to a change in the manner in which the alkali cations and their associated oxygen atoms are incorporated into the network. Uhlmann and Shaw[15] pointed out that while different investigators agreed about the presence of a minimum in the expansion coefficient for alkali borate glasses, there was a considerable divergence of opinion about the composition at which it occurred, the exact location ranging from about 15 to 30 mol % alkali, according to the particular set of results chosen. They also drew attention to the fact that the minimum is not observed when the expansion is measured in the molten state.[16] Their own measurements indicated that the thermal expansion coefficients of all simple alkali borates at first decrease with alkali content as far as 5–10 mol % alkali; they then remain substantially constant over the range 10–33 mol % alkali, after which there is a fairly steep rise in expansion coefficient. The coefficients also increased with the size of the cation from Li to Cs. The increase in expansion coefficient was found to begin at approximately the same composition at which nuclear magnetic resonance measurements indicated that the fraction of four-co-ordinate boron atoms was starting to deviate from the linear relationship with alkali content (Fig. 4.4). They concluded that the observed minimum in the expansion

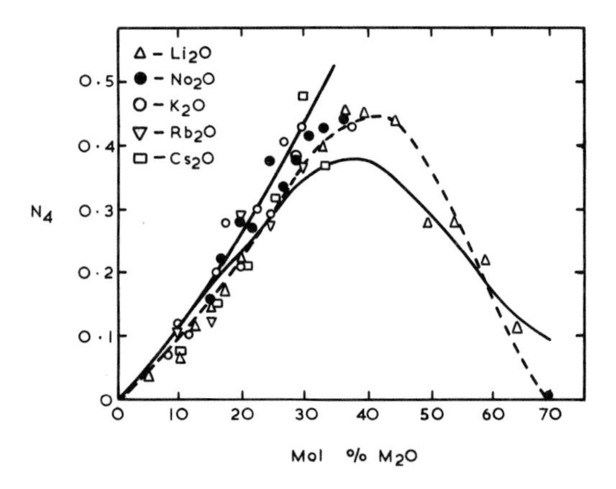

Fig. 4.4. Fraction of four-co-ordinate boron atoms (N_4) in alkali borate glasses as a function of alkali content.

coefficient was not a unique compositional effect, and may in fact be a consequence of liquid–liquid immiscibility, since all normally cooled alkali borate glasses with less than 20–25 mol % alkali are heterogeneous on a scale of a few thousand nm. They attributed the marked increase in expansion that occurs in the region of 30 mol % alkali to the presence of singly bonded oxygen atoms which reduce the effective connectivity of the network.

Borophosphate Glasses

Takahashi[17] found that the durability, i.e. resistance to hydrolysis by water, of simple sodium ultraphosphate glasses was markedly improved by the addition of boric oxide, and attributed this to the formation of BPO_4 groups. Boron phosphate itself is a four-connective network polymer, isostructural with silica (see pp. 130–131), and it would not be unreasonable to expect that a family of borophosphate polymers should exist, derived from boron phosphate in the same way that the silicate glasses are derived from silica. However, although homogeneous polymeric glasses are formed over the composition range 25–40 mol % boron phosphate in combination with soda, potash and lithia,[18] the structure of these polymers are quite different from the silicates and, as will be shown below, they must be regarded as being derived from networks of mixed three- and four-connectivity. Boron phosphate itself, unlike silica, does not form a glass and exists only as a crystalline polymer; but in compositions containing between 20 and 50 mol % alkali, homogeneous glassy polymers are formed over the entire range from pure phosphate to pure borate.[19]

Formation

Borophosphate polymers are readily produced by melting together phosphoric acid or ammonium dihydrogen phosphate, boric oxide and the appropriate quantities of alkali metal carbonates and alkaline earth metal oxides or carbonates. The temperatures required to effect this polycondensation are similar to those used for the production of ultraphosphate polymers (700°C), except when the proportion of boric oxide is 20 mol % or greater; these compositions require a higher temperature (up to 800°C) to complete the condensation.

Structure

Mazo and Navarro[20] found that the infra-red absorption spectra of alkali borophosphate glasses contain bands between 830 and 1000 cm^{-1} which, according to Müller,[21] are characteristic of tetrahedral BO_4 groups; but in this region of the spectrum there are also broad absorptions centred around 945 cm^{-1} due to the phosphate chain, and it is chiefly the absence of bands at 1440 cm^{-1} characteristic of trigonal BO_3 groups that leads to the conclusion that most of the boron atoms are four-co-ordinate. The evidence available from Raman spectroscopy is more positive.[22] The Raman spectrum of boric oxide, which contains only BO_3 groups, consists of just one major peak at 810 cm^{-1}; whereas the Raman spectrum of boron phosphate, in which all the boron atoms are four-co-ordinate, has no band at 810 cm^{-1}, but instead contains an intense, narrow band at

490 cm^{-1} which is easily distinguishable from any Raman bands that occur in the spectra of phosphates. Alkali borate glasses, which can contain both BO_3 and BO_4 groups, show Raman bands at both 810 and 490 cm^{-1}, and these change in intensity with alkali content in the expected way, the ratio of the intensity at 490 cm^{-1} to the total intensity of both bands increasing almost linearly with alkali content. The main features in the Raman spectra of borophosphate glasses are two strong bands at 690 and 1150 cm^{-1} that are associated with P—O vibrations, and two further bands at 490 and 810 cm^{-1} whose intensities vary with composition. An approximate relationship between the ratio of the Raman intensities at 490 and 810 cm^{-1}, and the proportions of three- and four-co-ordinate boron atoms which they represent, can be deduced from measurements of the corresponding intensities in the spectra of boron phosphate and of a

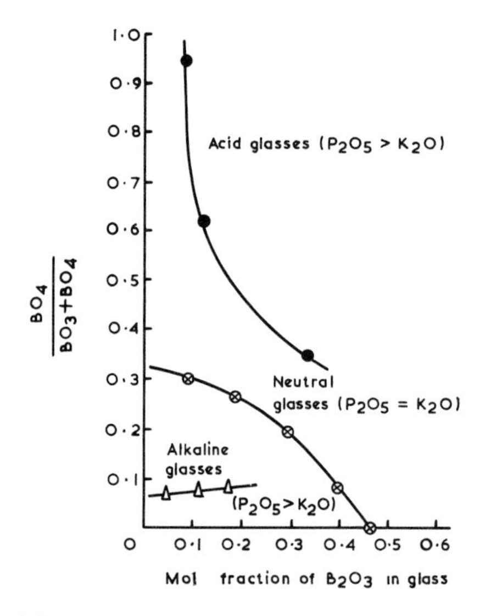

Fig. 4.5. Fraction of four-co-ordinate boron atoms in potassium borophosphate glasses as a function of boric oxide content. (From Ray, N. H. (1975). *Phys. Chem. Glasses,* **16** (4), 75. With permission.)

borophosphate glass of such a composition that the 490 cm^{-1} band is absent, using the main phosphate band at 1150 cm^{-1} as an internal standard. In this way the fractions of three- and four-coordinate boron atoms in borophosphate polymers of different compositions have been determined and the results, which are shown graphically in Fig. 4.5, fall into three distinct groups. Firstly, in compositions containing a stoichiometric excess of alkali over phosphoric oxide, practically all the boron atoms are present as trigonal BO_3 groups, the fraction of four-co-ordinate boron atoms being less than 0·1 for all glass-forming compositions in this range. Secondly, in compositions containing a stoichiometric excess of phosphoric oxide over alkali, when there is less than 10 mol % boric oxide in the glass, practically all the boron atoms are four-co-ordinate. As the proportion of boric oxide increases above 10 mol %, the fraction of four-co-ordinate boron atoms decreases rapidly towards one-third at 33 mol % boric oxide. Finally, in compositions containing equivalent proportions of alkali and phosphoric oxide—that is in metaphosphate compositions containing added boric oxide—the fraction of four-co-ordinate boron atoms decreases steadily with boron content from about one-third at very low concentrations of B_2O_3 to zero at about 47 mol % B_2O_3.

Properties

Before discussing the properties of borophosphate polymers it is necessary to establish a clear definition of cross-link density in these networks. In a borophosphate polymer containing a mole fraction x of phosphate (PO_4) groups and a mole fraction y of borate (BO_3 or BO_4) groups, the mole fraction of cations is $1\text{-}x\text{-}y$. For the purposes of calculation it can be assumed that the cations are monovalent, and that they will be preferentially combined first with any four-co-ordinate boron atoms in the network, next with the strongly acidic phosphate groups, and lastly with any non-bonding oxygen atoms in the BO_3 groups. Suppose the fraction of boron atoms that are four-co-ordinate is f. Then the mole fraction of four-co-ordinate boron atoms in the glass is fy, and the mole fraction of cations remaining to be shared between PO_4 and BO_3 groups is $1\text{-}x\text{-}y\text{-}fy$. So long as $x > 1\text{-}x\text{-}y\text{-}fy$, the mole fraction of phosphate groups that are linked to cations is $1\text{-}x\text{-}y\text{-}fy$. For values of $x \leqslant 1\text{-}x\text{-}y\text{-}fy$ the mole fraction of phosphate groups linked to a cation will be x (i.e. all the phosphate) and the mole fraction of BO_3 groups linked to a cation will be $1\text{-}2x\text{-}y\text{-}fy$. Then since each four-co-ordinate boron atom contributes two cross-linking points and each PO_4 or BO_3 group that is not linked to a cation contributes one, the number of cross-linking points in one "mole" of polymer is made up as in Table XIII.

Table XIII

For	$x > 1\text{-}x\text{-}y\text{-}fy$	$x \leqslant 1\text{-}x\text{-}y\text{-}fy$
BO_4 groups	$2fy$	$2fy$
BO_3 groups	$y(1\text{-}f)$	$y(1\text{-}f)\text{-}(1\text{-}2x\text{-}y\text{-}fy)$
PO_4 groups	$x\text{-}(1\text{-}x\text{-}y\text{-}fy)$	0
Total	$2x+2y(1+f)\text{-}1$	$2x+2y(1+f)\text{-}1$

Since the total number of network atoms in a "mole" or polymer is $x+y$, the cross-link density is

$$\frac{2x + 2y(1 + f) - 1}{x + y},$$

and this will be the maximum cross-link density for that composition. For given values of x and y the appropriate values of f can be obtained from Fig. 4.5. When there is a stoichiometric excess of phosphate, the cross-link density may be less than the maximum if the condensation has been interrupted before all the hydroxyl groups have been eliminated.

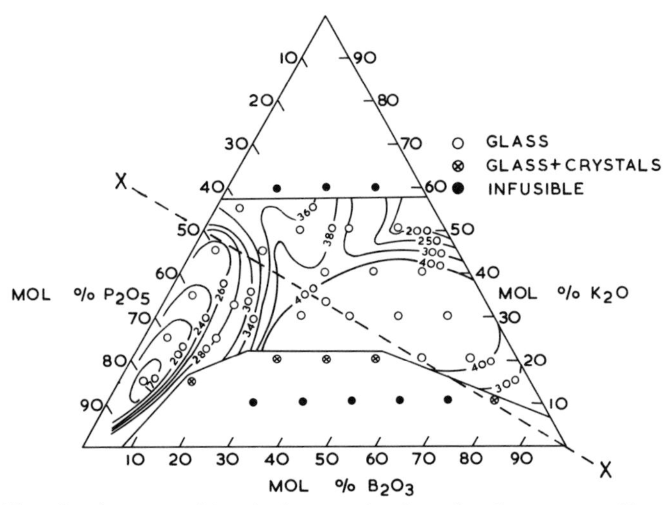

Fig. 4.6. Glass-forming compositions in the potassium borophosphate system with contour lines of glass transition temperature. (From Ray, N. H., Plaisted, R. J. and Robinson, W. D. (1976). *Glass Technology,* **17**, 66. With permission.)

Glass Transition Temperature

Glass transition temperatures in the three-component system $K_2O—B_2O_3—P_2O_5$ are shown as contour lines in the triangular diagram, Fig. 4.6. A cross-section of the contours along line X–X, that is for polymers having the composition xB_2O_3, $(1\text{-}x)K_2P_2O_6$ where x lies between 0 and 1, is shown in Fig. 4.7. For these compositions, which can be regarded as copolymers of boric oxide and potassium poly-(meta-)phosphate, the cross-link density can be calculated quite easily, as follows:

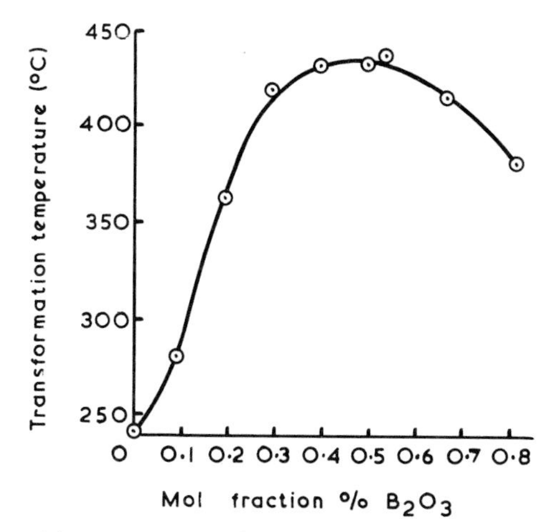

Fig. 4.7. Glass transition temperatures of potassium polyphosphate-boric oxide copolymers as a function of boric oxide content (cross-section of Fig. 4.6 along the line X–X. From Ray, N. H. (1975). *Phys. Chem. Glasses*, **16** (4), 75. With permission.)

The fraction of network-forming atoms that are boron atoms is $x/[x+\frac{1}{2}(1\text{-}x)]$. Each trivalent boron atom contributes one cross-linking point, but each four-co-ordinate boron contributes three—two at the boron atom and an additional one at a phosphorus atom that has lost its associated cation to the four-co-ordinate boron atom (Fig. 4.8). If the fraction of boron atoms that are four-co-ordinate is f, the fraction of network atoms that are cross-linking points (that is the cross-link density) is

$$\frac{3fx+(1-f)x}{x+\frac{1}{2}(1-x)}=\frac{2x(1+2f)}{1+x}.$$

(The same result is obtained by substituting $y = 1\text{-}x\text{-}y$ in the more general expression derived above.)

Plotting the glass transition temperatures of the series of potassium borophosphate polymers that contain equimolar proportions of potash and phosphoric oxide against the cross-link density calculated in this way gives the points shown in Fig. 4.9, which, apart from the last one at 54 mol % boric oxide, are an excellent fit to the curve predicted by the Gibbs–di Marzio equation

$$T_x/T_0=1/(1-Kx),$$

where x is the cross-link density and T_x, T_0 are the glass transition temperatures (in ° K) of the cross-linked polymer and the parent linear chain respectively; for this system the best fit is obtained with $K = 0{\cdot}43$, which is significantly lower

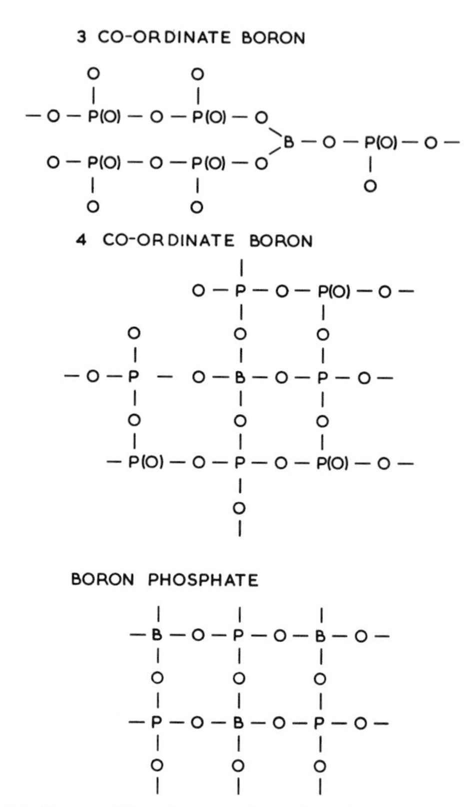

Fig. 4.8. Types of bonding in a borophosphate network.

than the value found for multicomponent ultraphosphate glasses containing no boric oxide.[12,13]

The same kind of treatment can be extended to borophosphate polymers containing different proportions of alkali metal cations and phosphoric oxide. For the "acidic" glasses, that is the systems in which there is a stoichiometric excess of phosphoric oxide over alkali cations, the glass transition temperature of any chosen composition increases with continued condensation and removal of hydroxyl groups in the same way as it does with simple ultraphosphate glasses, and for boric oxide contents between zero and about 10 mol %, the same relationship between glass transition temperature and calculated cross-link density holds good. Above 10 mol % boric oxide, the fraction of boron atoms that are four-co-ordinate changes too rapidly with composition (Fig. 4.5) to be able to assign precise values to the cross-link density, and in the region of compositions where there is a stoichiometric excess of alkali over phosphoric

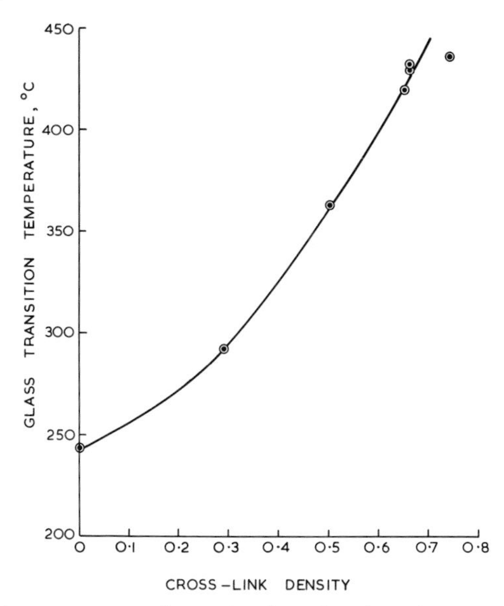

Fig. 4.9. Glass transition temperature of potassium borophosphate glasses as a function of cross-link density; experimental points superimposed on a calculated curve.

oxide there are two few reliable measurements to test the relationship properly. Nevertheless we may conclude that over a substantial part of the composition range up to 50 mol % boric oxide, the relationship between glass transition temperature and cross-link density in this system is similar to that found for the system of ultraphosphate glasses, and fits the Gibbs–di Marzio equation within experimental error.

Durability

Although the simple alkali borophosphates are all dissolved fairly easily by water, in the region of 80 mol % P_2O_5 and 5 mol % B_2O_3 there is a distinct minimum in the rate of solution.[19] This effect was also observed by Takahashi,[17] and it is noteworthy that this minimum in the rate of solution in water coincides with the compositions that contain the maximum fraction of four-coordinate boron atoms (Fig. 4.5).

Borophosphate glasses containing both alkali and alkaline earth metal cations are considerably more durable than simple ultraphosphate glasses of similar composition. The addition of 5 mol % of boric oxide to an ultraphosphate polymer containing (in mol %) 10 Li_2O, 10 Na_2O and 5 BaO increased its durability in running water by two orders of magnitude at the same glass transition temperature, and a number of multicomponent borophosphate glasses have been prepared which are as durable at ambient temperatures as most ordinary silicate glasses. One such composition contains (in mol %) 55 P_2O_5, 5 B_2O_3, 9 MgO, 9 CaO, 7 Li_2O and 15 Na_2O. Its glass transition temperature is 330°C, and the rate of solution in running water at 20°C is 5 μm $year^{-1}$, which is the same as that of a reasonably durable soda–lime–silica glass such as S95.[25] The rate of attack by water on the borophosphate glass, however, increases much more rapidly with temperature than it does for silicate glasses, the activation energy (85 kJ mol^{-1}) being about twice as great as it is for typical soda–lime–silica glasses. For this reason, measurements of durability made by the traditional methods of glass technology, which involve autoclaving samples of glass in superheated water at 130°C, give misleading results for borophosphate glasses, because very few of their practical applications involve exposure to temperatures above 30°C, and virtually none involve exposure to water above 100°C.

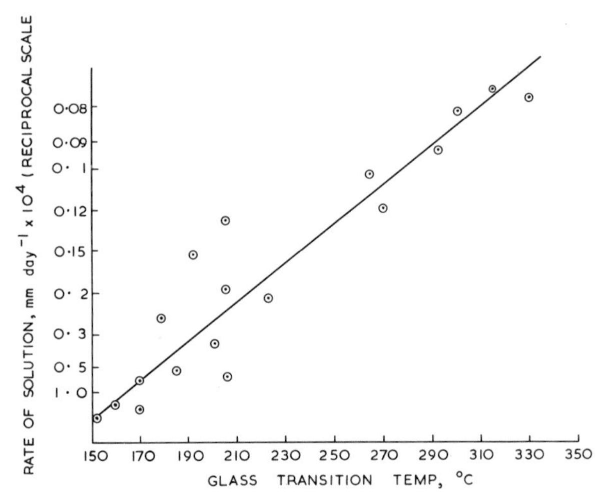

Fig. 4.10. Durability (as reciprocal rate of solution in water) of typical borophosphate glasses plotted against glass transition temperature.

As with ultraphosphate glasses the durability of borophosphate glasses increases with their glass transition temperature, the reciprocal of the rate of solution in water being roughly proportional to glass transition temperature. Experimental values for a number of different borophosphate compositions are shown in Fig. 4.10, which shows that reasonably durable glasses (rates of solution about 5–10 μm $year^{-1}$) can be obtained with glass transition temperatures as low as 200°C; this means that such materials can be processed in much the same way as the more higher softening-point organic thermoplastics. Comparison with corresponding results obtained for the ultraphosphate glasses (Fig. 3.10) shows that for polymers of similar glass transition temperature, borophosphates generally are between one and two orders of magnitude more durable towards water (see Table XII, p. 65).

Melt Viscosity

The temperature dependence of viscosity of a few borophosphate polymers has been investigated in some detail;[19] the results obtained for one composition containing (in mol %) 70 P_2O_5, 5 B_2O_3, 10 Li_2O, 10 Na_2O, 5 BaO that had been condensed to various stages, giving samples with four different values of cross-link density, are shown in Fig. 4.11. The activation energy of viscous flow for

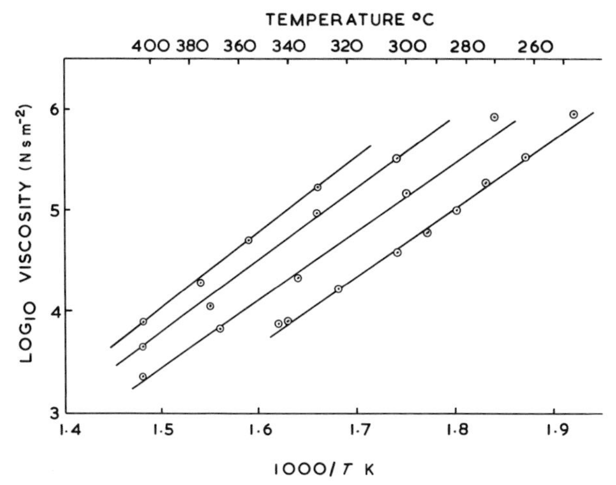

Fig. 4.11. Temperature dependence of melt viscosity for four borophosphate glasses of different cross-link densities. (From Ray, N. H., Plaisted, R. J. and Robinson, W. D. (1976). *Glass Technology*, **17**, 66. With permission.)

two borophosphate glasses was found to be 140 kJ mol^{-1}, which is significantly less than the values (190–200 kJ mol^{-1}) that have been found for ultraphosphate glasses. The activation energy is practically independent of cross-link density, but the differences between the temperatures at which the viscosity attains a value of 10^5 N s m^{-2} and the glass transition temperatures increase with cross-link density, showing that the glass transition temperature is not an isoviscous temperature for these polymers. It is significant that the activation energy of viscous flow in borophosphate polymers is much lower than it is for either of the parent network polymers (phosphoric oxide, 173·5 kJ mol^{-1};[26] boric oxide, 170–300 kJ mol^{-1} according to temperature.[27] Since the temperatures concerned are not nearly high enough to involve complete dissociation of P—O or B—O bonds, the flow of these cross-linked polymers in the molten state must involve "bond-switching", that is the successive transfer of bonds between network-forming atoms and oxygen atoms from one oxygen atom to the next. The lower activation energy of this process in borophosphates shows that the presence of both boron and phosphorus atoms in the same polymeric network makes it easier for oxygen bonds to transfer from one atom to another. This is readily understood if it is assumed that some of the boron atoms are linked through the normally non-bonding oxygen atoms of the phosphate groups, forming B—O—P linkages of the type known to be present in compounds such as Cl_3BOPCl_3, formed by the addition of boron trichloride to phosphoryl trichloride,[28] and in boron phosphate itself. The low sublimation temperature of Cl_3BOPCl_3 and the reversible dissociation of BPO_4 at its melting point clearly illustrate the thermal lability of B—O—P bonds of this kind, and it seems likely that viscous flow in the borophosphate network could occur by interchange of oxygen atoms between borate and phosphate groups as suggested in Fig. 4.12.

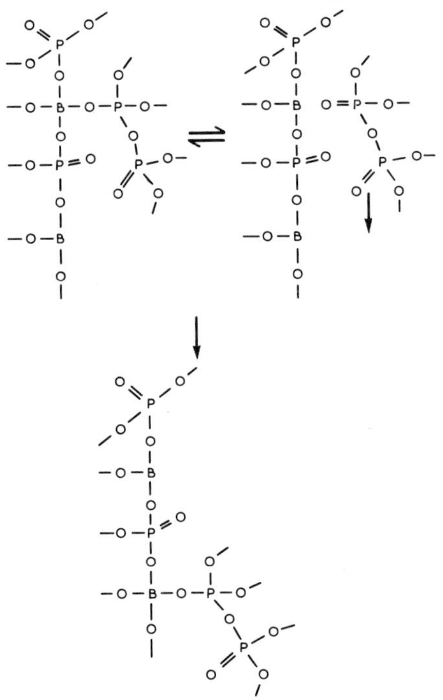

Fig. 4.12. A possible mechanism for viscous flow in a borophosphate network by bond interchange.

Surface Properties

An interesting and potentially valuable property of borophosphate polymers is their ability to resist surface contamination. Ordinary silicate glass that is exposed to the atmosphere rapidly acquires a surface film that is hydrophobic. Because of this, condensing water vapour forms a layer of minute droplets, so that spectacle lenses, for example, mist over in humid atmospheres when their temperature is below the dew-point. If the glass is rigorously cleaned with chromic acid, thoroughly rinsed in distilled water and dried in perfectly clean air, this does not happen; for a short time at least water vapour will condense as a thin, uniform film that is nearly invisible except by light that is incident at the correct angle to show interference colours. In most environments, however, this behaviour is short-lived and the glass soon returns to its normal hydrophobic condition, and mists up. Application of a thin film of a surfactant, such as a soap, will make ordinary glass non-misting but the effect is only temporary because the film is too easily wiped off.

Borophosphate glass, on the other hand, is inherently non-misting, and water vapour normally condenses as a thin film on its surface without any special pretreatment.[25] When the surface has been thoroughly washed with water and dried, the non-misting behaviour is temporarily lost, but returns spontaneously on standing in a normally humid atmosphere at ambient temperature for a short time. These glasses are therefore ideal for the manufacture of spectacle and other lenses which will not mist over in humid atmospheres.

Borophosphate glasses are also able to resist surface contamination of other kinds; for example they can be used as anti-scaling coatings to inhibit the deposition of crystalline scale from supersaturated aqueous solutions, and they also prevent the adhesion and growth of marine organisms on surfaces submerged for long periods in the sea.[29] The mechanism by which the surface of these polymeric glasses resist contamination and remain free from misting by condensed water vapour is not yet known with certainty, but it appears to be related to the way in which they react with water, whether as water vapour in the atmosphere or as liquid water, so as to produce an extremely thin film of mobile (i.e. partially hydrolysed) polymer on the surface. This film most probably consists of linear or almost linear fragments of the network that have become detached by hydrolysis of P—O—P or P—O—B bonds, and is sufficiently fluid, even at ambient temperatures, to spread over the surface of the glass and prevent the adhesion of crystals or marine organisms. It is also permanently hydrophilic, so that water condensing on to it from the vapour always forms a continuous film. Thorough washing will temporarily remove the partially hydrolysed polymer, but it will always be regenerated on standing in the presence of water or water vapour.

References

1. Imaoka, M. (1962). "Advances in Glass Technology: Technical Papers of the VIth International Congress of Glass, Washington, 1962", p. 366. Plenum Press, New York.
2. Stanworth, J. E. (1946). *J. Soc. Glass Tech.* **30**, 54T.
3. Leventhal, M. and Bray, P. J. (1965). *Phys. Chem. Glasses,* **6**, 113.
4. Parks G. S. and Spaght, M. B. (1934). *J. Phys. Chem.* **38**, 103.
5. Bray, P. J. and O'Keefe, J. G. (1963). *Phys. Chem. Glasses,* **4**, 37.
6. Jellyman, P. E. and Proctor, J. P. (1955). *Trans, Soc. Glass Tech.* **39**, 191T.
7. Wong J. and Angell, C. A. (1971). *Appl. Spectrosc. Rev.* **4** (2), 155.
8. Kim, K. S. and Bray, P. J. (1974). *Phys. Chem. Glasses* **15**, 47.
9. Bray, P. J., Leventhal, M. and Hooper, H. O. (1963). *Phys. Chem. Glasses,* **4**, 47.
10. Abe, T. (1952). *J. Amer. Ceram. Soc.* **35**, 284.
11. Huggins, M. L. and Abe, T. (1957). *J. Amer. Ceram. Soc.* **40**, 287.
12. Block, S. and Piermarini, G. J. (1964). *Phys. Chem. Glasses,* **5**, 138.
13. Krogh-Moe, J. (1965). *Phys. Chem. Glasses,* **6**, 46.
14. Krogh-Moe, J. (1960). *Phys. Chem. Glasses,* **1**, 26: Acta Cryst. 1960, **13**, 889.
15. Uhlmann, D. R. and Shaw, R. R. (1969). *J. Non-cryst. Solids,* **1**, 347.
16. Shartsis, L., Capps, W. and Spinner, S. (1953). *J. Amer. Ceram. Soc.* **36**, 35.
17. Takahashi, K. (1962). "Advances in Glass Technology: Technical Papers of the VIth International Congress on Glass, Washington 1962", p. 366. Plenum Press, New York.
18. Tien, T. Y. and Hummel, F. A. (1961). *J. Amer. Ceram. Soc.* **44**, 390.
19. Ray, N. H., Plaisted, R. J. and Robinson, W. D. (1976). *Glass Technology,* **17**, 66.
20. Mazo, J. L. and Navarro, J. F. (1970). *Boln. Soc. esp. Cerám.* **9**(6), 721.
21. Müller, K-P. (1969). *Glastech. Ber.* **42** (3), 83
22. Ray, N. H. (1975). *Phys. Chem. Glasses,* **16** (4), 75.
23. Ray, N. H., Lewis, C. J., Laycock, J. N. C. and Robinson, W. D. (1973). *Glass Technology,* **14**, 50.
24. Ray, N. H., Laycock, J. N. C. and Robinson, W. D. (1973). *Glass Technology,* **14**, 55.
25. British patent 1404914 to Imperial Chemical Industries.
26. Cormia, R. L., Mackenzie, J. D. and Turnbull, D. (1963). *J. Appl. Phys.* **38**, 2245.
27. Mackenzie, J. D. (1960). "Modern Aspects of the Vitreous State", Vol. 1 pp. 188–191. Butterworths, London.
28. Gerrard, W., Frazer, M. J. and Patel, J. K. (1960). *J. Chem. Soc.* 726.
29. U.K. patent application 48761/76 (1976) by Imperial Chemical Industries.

5

Four-connective Network Polymers

AMORPHOUS FOUR-CONNECTIVE NETWORK POLYMERS

Vitreous Silica and Silicate Glasses

Since there are already many excellent textbooks of glass technology[1, 2, 3] it would be superfluous, as well as out of place, to attempt an exhaustive treatment of silicate glasses in this volume. Instead it is proposed to discuss only those aspects of silicate glass structure and properties that will help to show their relationships to other kinds of inorganic polymers.

Structure

The common structural unit of the silicate glasses, as well as of all the crystalline silicates, is the tetrahedral SiO_4 group. It has unfortunately become customary to depict the structures of silicates with symbols—whether alphabetic or ideographic—that are roughly the same size, whether they represent silicon, oxygen, or other atoms. This convention gives no visual impression of the relative contributions of the different atoms to the packing of the network, the proportion of free space in the structure, or the molecular texture. If the atoms in vitreous silica, for example, are depicted approximately to scale as in Fig. 5.1, two general observations about the structure can be appreciated: firstly, that the material of this polymer is mostly oxygen; and secondly, that there is a good deal of free space in the structure. These observations can be put onto a more quantitative basis in the following way: the weight of oxygen contained in unit volume of various solids can be calculated from their elementary compositions

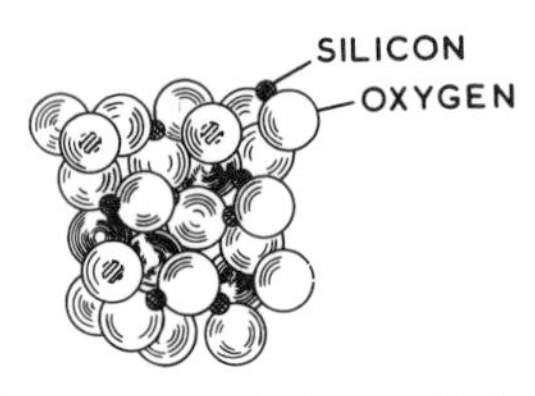

Fig. 5.1. Vitreous silica, approximately to scale. (From Holliday, L. (1975). "Ionic Polymers", Ch. 8. Applied Science Publishers, London. With permission.)

and densities—for example vitreous silica, which contains 53·33% oxygen by weight, has a density of 2·203 g cm^{-3}—consequently there are 1.175 g oxygen in each cubic centimetre. The results of some similar calculations for a number of different kinds of solid oxides, including oxygen itself in the solid state, are given in the third column of Table XIV. Such calculations show that, of all compounds, beryllium oxide contains the greatest weight of oxygen in unit volume.

Table XIV
Oxygen density and packing fraction in various solid oxides

Substance	Weight % oxygen	Density g cm^{-3}	Oxygen density g cm^{-3}	Oxygen packing fraction[a]
Glasses				
Vitreous silica	53·33	2·203	1·175	0·45
Soda-lime-silicate glasses	43·6–48·9	2·2–2·7	0·960–1·312	0·36–0·50
Crystals				
Quartz	53·33	2·655	1·416	0·54
Magnesium oxide	39·69	3.58	1·421	0·55
Oxygen at −253°C	100	1·426	1·426	
Titanium dioxide	40·1	4·26	1·706	0·66
Corundum (Al_2O_3)	47·1	3·99	1·879	0·72
Beryllium oxide	54·0	3·01	1·926	0·74[b]

[a] On the assumption that the oxygen atoms in the compound in question are the same size as in BeO.
[b] See text.

From X-ray diffraction studies it is known that the oxygen atoms in beryllium oxide are arranged in hexagonal close packing, and since no other oxide contains a higher density of oxygen, it might be reasonable to conclude that the oxygen atoms in beryllium oxide are as close together as they can ever be. From the geometry of a hexagonal close-packed assembly of spheres, it follows that 74%

of the volume of a crystal of beryllium oxide is occupied by oxygen atoms. Starting from this assumption, the fraction of the total volume that is occupied by oxygen atoms in any other oxide can be estimated by comparing its oxygen density with that of beryllium oxide, and some results obtained in this way are given in the last column of Table XIV. These estimates are only valid if the oxygen atoms in the compounds concerned are about the same size as they are in BeO, and this is obviously untrue in the case of molecular oxygen, for example. For silica, however, the estimate cannot be far wrong because on Pauling's scale[4] the electronegativity values of Be and Si are similar (1·5 and 1·8 respectively), showing that the Be—O and Si—O bonds both have about the same (*c.* 50%) fraction of ionic character. The values obtained for quartz, vitreous silica, and some representative silicate glasses show that in these materials the volume fraction occupied by oxygen is only from one-half to three-fifths of the maximum possible, and this result clearly confirms the general impression created by Fig. 5.1.

The relatively loose packing of the silica network has important consequences in terms of its physical properties, principally its compressibility and permeability. When vitreous silica is subjected to uniform hydrostatic pressure in excess of 70 kbar, its density is permanently increased although no crystallization occurs,[5] showing that the normal structure can be compacted to a new, metastable form. Crystalline silica can also be obtained in unusually dense modifications (see pp. 122–124) when it is formed under very high pressures.

The diffusion of gases through vitreous silica and silicate glasses has been the subject of many investigations, which have been reviewed by Doremus.[6] The rate of diffusion of hydrogen through vitreous silica, for example, is directly proportional to the partial pressure, showing that the gas diffuses through the silica network as hydrogen molecules. The solubility of gases such as helium, neon, argon and oxygen in vitreous silica has been discussed by Scholze[7] in relation to the number and sizes of the holes in the network, but he was unable to reach any quantitative conclusions about the relationships between gas solubility and structure because of insufficient knowledge of the glass structure. Nevertheless, it is clear from results such as these that the structure of vitreous silica must contain randomly distributed pores or holes of various sizes up to at least 0·3 nm in diameter. The crystalline modifications of silica, cristobalite and tridymite must likewise contain channels at least large enough to admit atoms of neon (0·23 nm in diameter). In the light of these observations, it is interesting briefly to review the available evidence about the structures of vitreous silica and the silicate glasses generally before discussing the properties of particular systems in detail.

The structure of the silicate glasses has been the subject of innumerable studies and some controversy for several decades, starting with Zachariasen's suggestion[8] that all oxide glasses have a random network structure built up from the same basic units that occur in the corresponding crystalline oxides. Thus vitreous silica, for example, is to be regarded as a giant molecule made up of SiO_4 tetrahedra joined at their vertices by sharing oxygen atoms to form a random three-dimensional network. The main difference between crystalline and vitreous silica according to this hypothesis is that in the glass the relative orientation of adjacent SiO_4 tetrahedra is infinitely variable, while in the crystal it is constant throughout the structure. This is equivalent to saying that in the glass there is a continuous distribution of Si—O—Si bond angles about some mean value. The advent of X-ray diffraction methods gave rise to a considerable number of investigations using the techniques of low-angle scattering and radial distribution functions, beginning with Warren's work[(9)] which gave much support to Zachariasen's theory. More recently the detailed interpretation of radial distribution functions, calculated from measurements of both X-ray and neutron scattering intensities, has resulted in two different kinds of structural models being proposed, one being the purely random network, while the alternative is a structure derived by distortion of a crystalline lattice. The presence of comparatively sharp peaks in the X-ray correlation function extending out to 1·4 nm,[(10)] and of anisotropy in small areas revealed by high-resolution transmission electron microscopy,[(11)] has been interpreted as indicating the existence of ordered regions in glasses of the order of 1 nm in diameter.[(12)] To account for these observations an alternative to the totally random network theory appeared to be necessary, which is that the structure can be regarded as being like a grossly disordered crystal, and partly retains the structural order present in the crystal, at least over very small regions.

Deciding which of these alternatives best fits all of the observations is none too easy. The results of X-ray diffraction and neutron scattering experiments give only one-dimensional representations of a three-dimensional structure, and these are averaged over the whole sample. Consequently, even if it were possible to determine radial distribution functions with complete certainty, the structure would not be uniquely defined. As a result, the only way to differentiate between alternative hypotheses is first to postulate a model structure, and then by calculation of its radial distribution function to test how well it agrees with experimental observations. Evans and King,[(13)] and Bell and Dean[(14)] have constructed physical models of vitreous silica on the basis of a random network structure, and have calculated from actual measurements on the models their radial distribution functions and other parameters, the most important being density and the distributions of Si—O—Si bond angles. A recent analysis by Da Silva and others[(15)] of Mozzi and Warren's experimental data[(16)] gave 152° as the most probable Si—O—Si angle, and this is in excellent agreement with Bell and Dean's model, for which it was found that to reproduce the experimental correlation function the mean value of the bond angle in the model had to be 153°. Bell and Dean also computed the infra-red and Raman frequencies to be expected from their model[17] and the predicted frequencies showed good agreement with the actual vibrational spectrum of vitreous silica. Building up physical models of random structures is necessarily very tedious, and an alternative approach that has been adopted is to construct models with a computer. Models based on a variety of crystalline polymorphs have been used to interpret experimental correlation functions by Fourier transformation of appropriately broadened powder diffraction patterns. This approach has been used for vitreous silica and germania by Konnert and Karle[10, 18] who concluded that in SiO_2 glass the short-range order was similar to that in tridymite and extends to distances of 1·5–2·0 nm, much longer than previously reported. In this respect their model differs from that of Bell and Dean, because the structures of both tridymite and α-cristobalite contain six-membered rings of SiO_4 tetrahedra, while Bell and Dean's model contains a much larger number of four- and five-membered rings. An objective assessment of the relative merits of the random network theory and the "crystallite" theory has been provided by Wright and Leadbetter,[19] who concluded that they represent two extremes, with the truth for most systems lying somewhere in between.

Alkali Silicates

Starting from the assumption of a four-connected random network structure for vitreous silica, the addition of metal cations should cause a gradual reduction in cross-link density accompanied by a progressive decrease in glass transition temperature and melt viscosity. With a uniform distribution of cations, the composition at which the cross-link density reaches zero, so that only linear and simply branched segments of network remain, corresponds to that of the metasilicate (i.e. 50 mol % SiO_2). In reality, however, the properties of the simple alkali silicate glasses do not change smoothly and progressively with increasing alkali content from those of pure silica to those of the alkali metasilicate; and neither do the properties of the alkali metasilicates when examined in detail agree with what would be expected of high molecular weight linear polymers. For these reasons, the water-soluble alkali metal silicates, which already have a long history as important industrial products, should also be of considerable interest to polymer scientists. Because so few soluble inorganic polymers are available for study it might be expected that the alkali silicates would be a useful source of information about the properties of silicates generally from their solution properties, but unfortunately this is not the case. It is true that the two commonest alkali metal silicates both form a continuous series of homogeneous mixtures with water ranging from solid, partially hydrated glasses, through highly viscous liquids, to dilute and comparatively mobile aqueous solutions; but the physical properties of these systems show no simple dependence on composition or concentration, in fact within certain composition ranges they exhibit such rapid changes with concentration that it seems certain that more than one kind of polymer must be involved.

The dissolution of an anhydrous alkali silicate glass in water is a complex process involving both chemical change as well as physical solution. The solution first formed in contact with the solid glass generally contains different proportions of alkali cations and silica from those present in the glass, and even the apparently simple process of diluting an alkali silicate solution with water initiates slow changes in pH, electrical conductivity, specific viscosity and other colligative properties, which continue to alter towards their equilibrium values during the course of hours or even days.[20] Consequently it is not feasible to infer the structure of solid alkali silicates from the properties of their aqueous solutions. It will suffice to discuss two properties which are in direct conflict with what is known about the solid glasses and their properties in the melt. Firstly, lithium silicate glasses made by fusion of lithium carbonate and silica cannot be dissolved in water to form concentrated, viscous solutions like those that are readily obtained by treating sodium or potassium glasses with hot water or steam. Nevertheless, lithium silicate solutions can be prepared without difficulty by dissolving freshly precipitated silica in concentrated lithium hydroxide solutions, and made in this way their properties are very similar to solutions of sodium and potassium silicates.

Secondly, the molecular weight of sodium silicates in aqueous solution, as measured either by viscosity[21] or by light-scattering[22] is much lower than their stoichiometry would suggest: at the metasilicate composition (50 mol % silica) the weight average molecular weight is only 71, which corresponds to complete dissociation into discrete $SiO^{2}_{3}{}^{-}$ anions and Na^+ cations. As the ratio of SiO_2 to Na_2O is increased, the average molecular weight increases (Fig. 5.2), but even at the composition $Na_2O \cdot 3\ SiO_2$ the weight average molecular weight is only 300, corresponding to an average chain length of no more than 3 to 4 SiO_4 units.[23] The molecular weight of potassium silicate at similar compositions is somewhat higher, but still only corresponds to chains of 4–5 silicate units. In more concentrated solutions (up to 30–50% by weight) the viscosity at constant concentration passes through a pronounced minimum at the disilicate composition $Na_2O \cdot 2\ SiO_2$,[24] rising steeply with increasing silica content and more gradually with increasing alkali content. These results show clearly that the processes of dissolving an alkali silicate glass in water and subsequently diluting the initial solution with additional water cause fundamental alterations in the polymer, probably involving the hydrolysis of Si—O—Si bonds and the hydration of the fragments produced. The extent of hydrolysis evidently varies both with concentration and with composition (i.e. Si : Na ratio), so that the apparent molecular weight varies with both concentration and composition. For these reasons, it is only from studies of the anhydrous alkali silicates that reliable information can be obtained about their polymeric properties and structure.

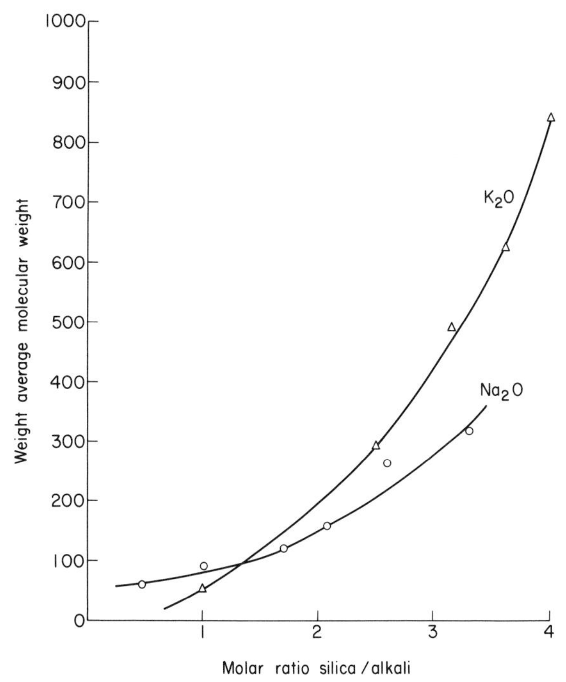

Fig. 5.2. Weight average molecular weights of alkali silicates in aqueous solution as a function of alkali content. (From Holliday, L. (1975). "Ionic Polymers", Ch. 8. Applied Science Publishers, London. With permission.)

Properties: Melt Viscosity and Glass Transition Temperature

The glass transition temperature of silica is extremely sensitive to water content, so that values reported in the earlier literature are variable; the most reliable value appears to be 1200°C[25, 7] for anhydrous silica, falling to 1075°C for 0·25 mole % H_2O. The introduction of even quite small proportions of alkali metal oxides produces a steep drop in glass transition temperature and melt viscosity; for example 2·5 mol % potash lowers the viscosity of molten silica from 2×10^6 N s m^{-2} at 1700°C to 200 N s m^{-2},[26] and other alkali metal and alkaline earth metal oxides produce similar effects. As the alkali metal content is increased, however, the change in viscosity with alkali concentration rapidly becomes smaller (Fig. 5.3). The viscosity of any alkali silicate of a given composition varies with temperature according to an equation of the form

$$\eta = A \exp(E/RT),$$

but the values of A and E vary considerably with silica: alkali ratio. With increasing alkali content, the value of E decreases rapidly from 585 KJ mol^{-1}

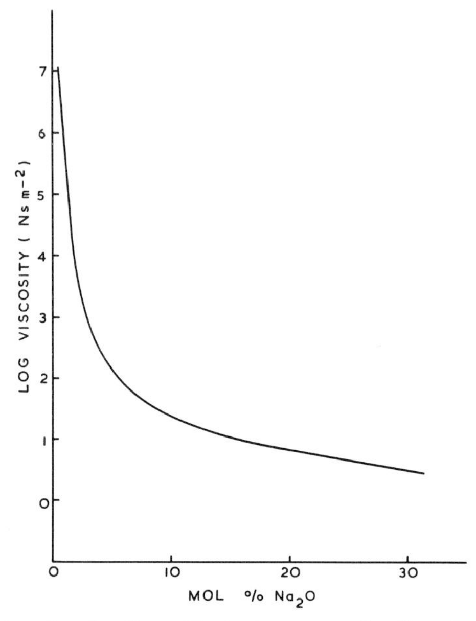

Fig. 5.3. Melt viscosity of sodium silicate glasses as a function of alkali content.

down to 200 KJ mol^{-1} at 10 mol% alkali oxide, after which it falls more slowly to 85 KJ mol^{-1} at 50 mol% alkali (Fig. 5.4). Similar changes are observed in the value of A.[27]

Coefficient of Expansion

The thermal expansion coefficients of all the alkali silicates begin to increase rapidly with alkali content at 12 mol% alkali. Below this concentration, the expansion coefficients are very low and nearly the same as that of vitreous silica (Fig. 5.5) A similar behaviour has been observed for the barium silicates. At compositions containing more than 20 mol% alkali oxide, the thermal expansions increase in the order Mg, Ca, Sr, Ba, Li, Na, K.

Structure

To be satisfactory, a structural model of the alkali silicates must account for the rapid changes in the behaviour of their properties with composition in the region of 10 mol % alkali, as well as the more gradual changes at higher concentrations. The comparatively low melt viscosity of the alkali metasilicates (less than 1 N s m^{-2} at the melting temperature of the crystalline phase) shows in the liquid

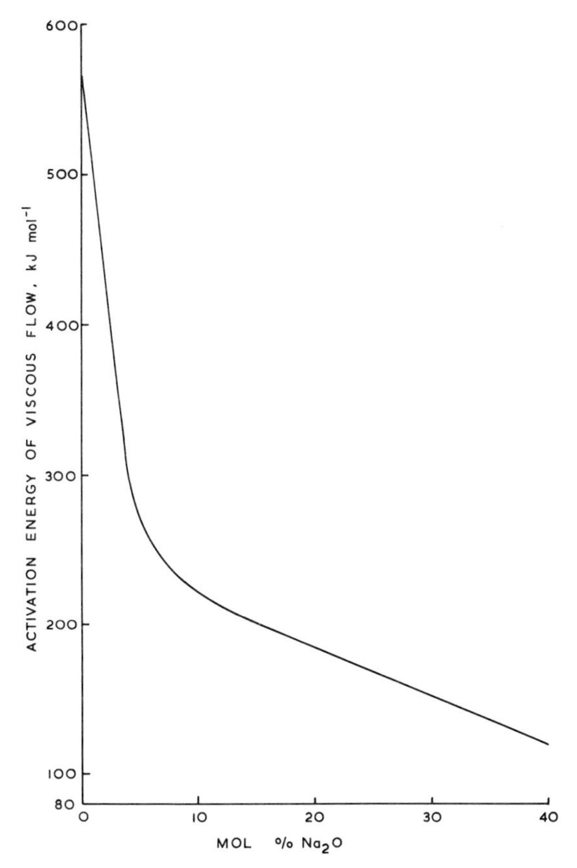

Fig. 5.4. Activation energy of viscous flow for sodium silicate glasses as a function of alkali content.

state that they are almost certainly not linear polymers of high molecular weight as their stoichiometry seems to indicate. Furthermore, the melt viscosities of all the alkali silicates, including the metasilicate, are practically independent of shear rate over a wide range of both composition and temperature, showing that their chain lengths in the molten state are below the critical value for non-Newtonian flow (see Chapter 3, p. 66). This suggests that viscous flow in these polymers probably involves bond-switching even at the metasilicate composition.

A structural model proposed by Endell and Hellbrügge[28] postulated a progressive breakdown of the silica network as more and more alkali is added, giving at first a mixture of two-dimensional sheet anions analogous to those in the naturally occurring layer silicates such as talc and mica; then as further alkali is added a mixture of long-chain and cyclic molecules is formed, and finally when alkali in excess of the metasilicate composition is present, there is a progressive decrease in chain length down to the orthosilicate composition. This model is unsatisfactory for two reasons; firstly, it fails to account for the sudden

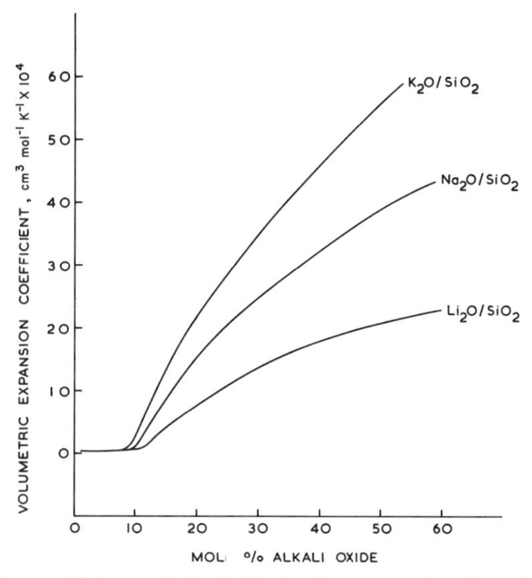

Fig. 5.5. Expansion coefficients of alkali silicate glasses as a function of alkali content.

changes in slope of the viscosity-composition and thermal expansion-composition curves in the region of 10–12 mol % alkali; and secondly, it requires that the metasilicate (50 mol % alkali) should be a long-chain polymer, which, as already mentioned, is unlikely to be the case in view of its low viscosity and purely Newtonian flow behaviour.

A purely random breakdown of the silica network into fragments of gradually decreasing connectivity, as proposed by Douglas[29] is equally unsatisfactory, because this would predict a smooth variation of physical properties with composition, and a glass transition temperature that decreased gradually with increasing alkali content according to the decrease in cross-link density; it cannot account for the sudden drop in melt viscosity and glass transition temperature that is observed over the composition range from 0–10 mol % alkali.

Bockris and others [30, 31] postulated a more satisfactory model, in which it is assumed that the melt consists of condensed rings built up of $Si_3O_9^{6-}$ and $Si_4O_{12}^{8-}$

anions, so that the molecules are always approximately spherical. At the metasilicate composition, the anions present are mainly $Si_3O_9{}^{6-}$ and $Si_4O_{12}{}^{8-}$; as the alkali content is reduced, these anions condense so that at 33 mol % alkali they are $Si_6O_{15}{}^{6-}$ and $Si_8O_{20}{}^{8-}$, and at 11 mol % alkali the average size of anion would be represented by $Si_{24}O_{51}{}^{6-}$ and $Si_{32}O_{68}{}^{8-}$. Bockris then suggests that when the alkali content is reduced still further, these very large fused-ring anions become unstable and are transformed into a three-dimensional network structure, so that round about 10 mol % alkali the structure suddenly changes from an aggregate of discrete, though complex, anions to a highly connected network. Below 10 mol % alkali the structure is a random network whose cross-link density increases with decreasing alkali content. This model accounts satisfactorily for the low melt viscosity of compositions in the neighbourhood of the metasilicate, and for the gradual rise in glass transition temperature and viscosity as the alkali content is reduced towards 10 mol %. The sudden increase in melt viscosity and glass transition temperature that is observed at about 10 mol % alkali can be ascribed to the change of structure into a random three-dimensional network. However, the model provides no explanation as to why this change occurs at the particular composition at which the rapid changes in physical properties are observed.

There is little in the way of structural evidence, either from X-ray analysis or vibrational spectroscopy that might help to explain these changes in properties of the alkali silicates at around 10 mol % alkali. Etchepare[(32)] examined the Raman spectra of lithium, sodium and potassium silicates of varying alkali content, but his results (which are discussed in greater detail below) do not extend to alkali contents below 15 mol %. Wilmot[(33)] recorded the Raman spectra of sodium silicate glasses covering the range 0–43 mol % Na_2O, but found no marked changes in the region of 9–13 mol % soda; Simon and McMahon[(34)] studied the infra-red absorption spectra of sodium silicate glasses over the same range, but found significant changes only when the alkali content was greater than 20 mol % alkali. Porai–Koshits[(35)] studied the way in which the X-ray diffraction pattern of sodium silicate glasses changes with composition. He found that the largest changes in the radial distribution curve occurred at 0·23 and 0·35–0·37 nm, corresponding to the Na—O and the Na—Na distances respectively. Guaker and Urnes also investigated these changes[(36)] and reached the conclusion that there must be a spectrum of Na—Na distances in these glasses with a range of 0·3 to 0·5 nm. This is consistent with Bockris's model, but the smallest proportion of alkali in the glasses investigated was 15 mol %, so that it is not possible from these results to reach any conclusion about structural changes in the region of 10–15 mol % alkali. Katsuki and Aoki[(37)] calculated the electronic states of non-bridging oxygen atoms in sodium disilicate glass by the self-consistent field method and showed that the total energy decreased abruptly when the oxygen–oxygen distance became less than 0·31 nm; this means that discrete molecules become stable above this point, and the system would be expected to exhibit the properties of an assembly of isolated small units. This supports Bockris's model in the region of the disilicate composition (33 mol % alkali) and explains the low viscosity in this region, but an explanation of the marked changes in behaviour of the melt viscosity and expansion coefficient with composition in the region of 10 mol % alkali is still required. Another, quite different, possibility is that these effects are due to a change in molecular texture, that is to say density fluctuations on a scale of one to ten nm, rather than atomic structure on a scale of tenths of a nm. Recent observations using low-angle X-ray scattering[38, 39] and high-resolution electron microscopy[40] have revealed the presence of inhomogeneity in the so-called "simple" alkali silicate glasses. The inhomogeneity can be intensified by heat-treatment, leading ultimately to phase separation. Porai-Koshits found that the temperature above which no opalescence was observable—that is the temperature of complete miscibility—reached a maximum at about 10 mol % Na_2O in sodium silicate glasses. Hammel[41] found that between 8·7 and 13·6 mol % Na_2O, heat-treatment resulted in a connected two-phase texture indicating spinodal decomposition, that is separation without nucleation.[42] These results indicate that the tendency towards phase separation is actually greatest in that region of composition where the melt viscosity and glass transition temperature start to change more rapidly with composition and it is therefore possible that the changes in the behaviour of these physical properties could be due to a change in texture resulting from an inversion of phases. Transmission electron micrographs of barium silicate glasses that have a region of liquid–liquid immiscibility centred around 10 mol % BaO show such a phase inversion very clearly as the barium oxide content is increased from 4 to 24 mol %.[43] The same phenomenon occurs in the alkali silicates,[44] but in this case the miscibility gaps are metastable—that is they occur at temperatures that are entirely sub-liquidus. When the composition lies on the silica-rich side of the inversion point, the polymer consists of a dispersion of alkali silicate droplets in a silica matrix; on passing through the inversion point to the alkali-rich side it changes to a dispersion of silica droplets in an alkali silicate matrix. Heat treatments of sodium silicate compositions in this region indicate that the droplet size increases with the square root of time.[45]

The existence of a two-phase microstructure in a polymer will affect many of its physical properties, and the effects on viscosity and glass transition temperature can be profound. The effect of phase separation on the viscosity of sodium silicate glasses has been extensively studied.[46] When the silica-rich phase is discontinuous, the melt viscosity is lower by several orders of magnitude than when the silica-rich phase is continuous, and the dilatometric softening point of a glass with an interconnected silica-rich phase is considerably higher than that of a similar composition with a continuous alkali-rich phase.[47]

Thus the abrupt changes in behaviour of the melt viscosity and glass transition temperature of the alkali silicates in the region of 10 mol % alkali could be the result of phase inversion of a two-phase system in this composition range; and since the system becomes a homogeneous liquid at temperatures above the liquidus, this explanation could be tested by investigating the behaviour of viscosity with composition at sufficiently high temperatures.

The thermal expansion of a two-phase solid is mainly dependent on the volume fractions of the two phases, and very little upon their morphology.[48] Consequently, phase inversion would not be expected to produce large changes in expansion coefficient, but the miscibility gaps in the alkali silicate systems are markedly asymmetric[49] and the volume fractions of the silica-rich and alkali-rich phases will change in a non-linear fashion with composition in the region of the miscibility gap.

It therefore seems likely that phase separation and phase inversion is at least partly responsible for the changes that occur in some physical properties of the alkali silicates when the alkali content is increased beyond 10 mol %.

Vibrational Spectra

The infra-red and Raman spectra of the alkali silicates have been recorded by Etchepare,[32] Jellyman and Procter,[50] Simon and McMahon,[34] Gross and Romanova[51] and Langenberg.[52] In vitreous silica there is a strong, characteristic infra-red absorption band at 1 100 cm^{-1} that is attributed to the Si—O—Si stretching vibration[53] and which does not appear in the Raman spectrum. This apparent obedience to the selection rules for centrosymmetric groups suggests that, so far as this vibrating unit is concerned, there must be some degree of order even in this amorphous polymer. In the alkali silicates, as the alkali content increases, the intensity of this band in the infra-red spectrum decreases and a corresponding band begins to appear in the Raman spectrum; at approximately 33% Na_2O, the Raman band at 1100 cm^{-1} is the most intense in the spectrum, while it has all but disappeared from the infra-red spectrum. When the concentration of soda exceeds 25% a new infra-red band appears at about 950 cm^{-1}, which at higher alkali contents splits into two bands at 940 and 1075 cm^{-1}. The fact that these bands occur at the same frequencies in lithium, sodium and potassium silicates shows that they must be attributed to vibrations of $Si—O^-$ or $O^-—Si—O^-$ groups. There is also a marked change in the Raman spectrum in the 500–600 cm^{-1} region when the alkali content exceeds 25 mol %. At this composition a broad, comparatively featureless band, that occupies most of the 200–500 cm^{-1} region in the composition range 0–25 mol % alkali, suddenly disappears and is replaced by a relatively narrow, strongly polarized and intense peak centred about 600 cm^{-1}. The same band is present in the spectra of all three alkali silicates and its intensity and position are practically independent of the nature of the cation. Wilmot[33] attributed this peak to a bending vibration of an Si—O⁻group. Another band at 955 cm^{-1}, which was very weak in compositions with less than 20 mol % alkali, became intense at 45 mol %. Etchepare[32] prefers to assign this latter band to a vibration of the Si—O^- group and the 600 cm^{-1} band to a bending mode of the O—Si—O unit.

Borosilicate Glasses and Glass-ceramics

The technical advantages of borosilicate glasses compared to glasses based on either silica or boric oxide alone were originally recognized in 1915 by Sullivan and Taylor,[54] who first formulated the range of borosilicate compositions that has come to be known as "Pyrex" glass. This is a family of four-connective inorganic network polymers which, in comparison with soda-lime-silicate glasses, have much lower expansion coefficients, and consequently greatly improved resistance to thermal shock, greater chemical durability, especially towards alkaline reagents, and very good working properties in the molten state. The distinguishing feature of the "Pyrex" group of glasses is the presence of about 12% by weight of boric oxide and around 80% by weight of silica (the original limits were from 20% B_2O_3, 70% SiO_2 to 5% B_2O_3, 90% SiO_2, but the majority of "Pyrex" type glasses made at present contain 79–81% SiO_2 and 10–13% B_2O_3) which corresponds to a molar ratio $SiO_2:B_2O_3 = 7{\cdot}78$ and an atomic ratio $Si:B = 3{\cdot}89$. The total metal oxide content is evidently 8% by weight which, assuming an average molecular weight of 60 for these oxides (chiefly Na_2O with minor amounts of MgO, CaO and Al_2O_3), corresponds to a molar ratio of network-forming elements to cations of about 11. An important question is whether "Pyrex"-type glasses are genuine copolymers of silica and boric oxide with randomly distributed cations, or a mixture of two different network polymers with the cations preferentially associated with one or other of the components.

It is a characteristic feature of polymers that they are generally immiscible in the liquid or plastic state.[55] Amongst organic polymers single-phase homogeneous mixtures are extremely rare, yet on superficial examination "Pyrex"-type glasses appear to be homogeneous, suggesting that they must be copolymers. However, inorganic glasses are normally cooled rather quickly through the molten region in relation to their high viscosities; even if phase separation did occur, the rate of increase of viscosity over the temperature range within which liquid–liquid separation is likely can be so great that the droplets of the separating phase do not have time to grow to an observable size before the material becomes a rigid solid. It is therefore possible for a glass that ought to be a two-phase system to appear perfectly homogeneous by visible light. Since the discovery of "Pyrex", many other borosilicate compositions have been studied, and it was subsequently discovered that one important class of glass compositions of this type have the characteristic property that they become opalescent on heating. These are the "Vycor" glasses which contain between 55–70% SiO_2 and from about 20–40% B_2O_3.[56] Electron microscopy shows that all glasses of this type are two-phase systems even before heat treatment, and that the effect of heating at temperatures near to the glass transition, that is in the range 500–600°C, is to cause the disperse phase to aggregate into droplets large enough to scatter visible light, thus causing the glass to become opalescent. Continued heat treatment within the region of immiscibility causes further growth of the disperse phase until eventually a network of interconnected regions of the separated phase is formed. In this condition it is possible selectively to leach out the disperse phase from the glass, because it is soluble in aqueous hydrochloric acid while the matrix, being almost pure silica, is unaffected. Leaching results in a porous residue of nearly pure silica glass that retains the original size and shape of the phase-separated glass. The pores that are obtainable by this process range in diameter from about 0·2 to 0.5 nm,[57] so that leached "Vycor" glass can be used for filtering bacteria and viruses; the pore volume may amount to 30% and the internal surface may be as great as 200 m^2g^{-1}, so that another possible application for this material is as a catalyst support. After drying, if the porous glass is carefully heated to 900–1200°C, it can be made to shrink uniformly into a non-porous, transparent piece of vitreous silica that retains the same geometrical form as the original glass. Its properties are close though not identical to those of fused quartz.[58]

It is evident from the behaviour of the "Vycor"-type glasses that they at least are mixtures of two polymers, one being not far removed from pure silica in composition, the other being an alkali borate glass. Rockett and others[59] studied the phase diagram for the system silica-sodium tetraborate and showed that there is an extensive region of liquid immiscibility below 800°C so that, at 500°C for example, two phases are present over the whole composition range from 10–95% SiO_2. Skatulla and co-workers[60] examined a series of borosilicate glasses with compositions lying on a tie-line in the ternary diagram joining pure SiO_2 to a binary sodium borate glass containing 16 mol % Na_2O (Fig. 5.6). They found that all such glasses with less than 70 mol % SiO_2 were two-phase systems, although without heat treatment none appeared opalescent. In another study of sodium borosilicate glasses containing between 60 and 80 mol % SiO_2 (13–16 mol % B_2O_3), Bokin and others[61] found that fibres drawn from glasses of this type developed worm-like structures on heating at 720°C, suggesting that phase separation had already occurred during the drawing of the fibre, and was emphasised by subsequent heat treatment. Zhdanov and co-workers have suggested that silica and boric oxide probably cannot form a common oxygen network (i.e. a random copolymer) because of the tendency of boric oxide to form chains of six-membered boroxole rings. The binary system B_2O_3—SiO_2 is difficult to study because the liquidus temperatures are high and the volatility of boric oxide is considerable at the temperatures involved; when minor amounts of alkali oxides are added, the cations appear to be associated preferentially with BO_4^- groups.[62] Thus the evidence seems to suggest that all the borosilicate glasses are either mixtures of two polymers, or possibly block copolymers with a

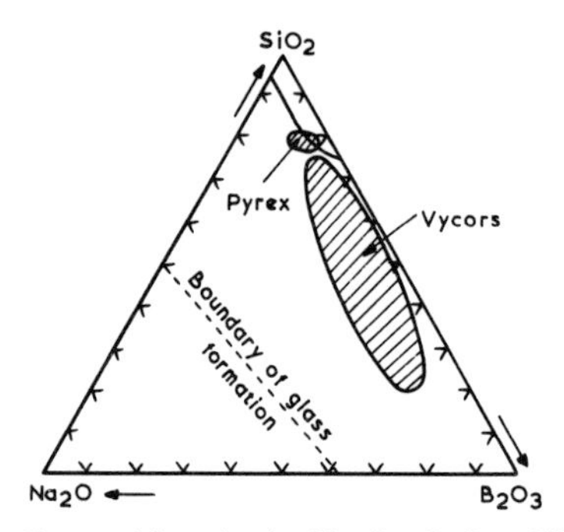

Fig. 5.6. Compositions in the Na_2O—B_2O_3—SiO_2 system.

strong tendency towards phase separation into a silica-rich matrix and a disperse phase of alkali borate glass. This explains their relatively low coefficient of expansion which will be mainly controlled by the volume fraction of the matrix, and their resistance to chemical attack which will be maintained as long as the more soluble disperse phase remains discontinuous.

Glass-ceramics

Glass-ceramics are polycrystalline solids prepared by the controlled crystallization of an inorganic glass, usually with a minor proportion of a residual glassy phase present. They represent an intermediate class of inorganic polymers lying between the amorphous silicates and the fully crystalline polymeric silicates, and because of their commercial importance they deserve a brief mention here. The first glass-ceramic was produced by Réaumur, in 1739,[63] who packed glass bottles into a mixture of sand and gypsum and heated them for several days at red-heat. The glass was converted into an opaque, porcelain-like material that more or less retained the original shape of the bottles, although with such relatively crude temperature control, some distortion was inevitable. Modern glass-ceramics are made by casting or otherwise shaping an article from a glass of specially chosen composition, and then subjecting it to a closely controlled heat treatment that causes a large fraction of the glass to crystallize without gross deformation of the article (some shrinkage is unavoidable and its extent is predictable). The types of glass that are particularly suitable for the production of glass-ceramics are magnesium aluminosilicates, lithium-magnesium silicates, and lithium-zinc silicates. Nucleation catalysts are added to initiate the crystallization process, and these are usually either titanium dioxide, phosphoric oxide and metal phosphates, or fine divided metals such as copper.[64] Effective nucleation is important because the aim is to produce an extremely fine-grained ceramic, the average crystallite size being only 100–500 nm. For this reason, comparatively large concentrations of nucleating agent have to be used, for example 10–12% by weight of titanium dioxide, or 3–6% by weight of phosphoric oxide. The optimum temperature for nucleation is above the glass transition temperature, and the temperature for maximum crystallization rate is higher still, so that the heating process must be very carefully controlled in order to avoid undue deformation of the article.

In a study of the formation of glass-ceramics in the magnesium aluminosilicate system with titanium dioxide as the nucleation catalyst, Maurer[65] found that even before heat treatment the intensity of light-scattering was three times greater than would be expected from a homogeneous glass; while at the lowest temperature of heating, the depolarization increased sixfold with only a 30% change in total scattering. He concluded that crystallization of an emulsion-like disperse-phase was taking place, and that the glass must have been a two-phase system to begin with, like the borosilicates discussed above. Titanium dioxide reacts in the melt with magnesium oxide to form a network polymer that is incompatible with the silicate network and separates on cooling as a second liquid phase. It is this finely dispersed phase which crystallizes first. Phosphoric oxide, another common nucleating agent, behaves in a similar way; high molecular weight metal polyphosphates that are formed in the melt separate out as a liquid disperse phase which subsequently crystallizes on re-heating the glass.[64] Thus an inportant step in the formation of a glass-ceramic is the production of a two-phase precursor glass containing two different, immiscible oxide networks. This is another example of the mutual incompatibility of high polymers. The properties of glass-ceramics are nowadays too well known to require detailed description. Because of the very high degree of crystallinity in these materials they behave in many respects more like the crystalline silicates than like the silicate glasses, for example they have much greater resistance to high temperatures, while the fineness and uniformity of the dispersion of crystallites ensure that they are essentially isotropic so far as mechanical properties are concerned.

Beryllium Fluoride and the Fluoroberyllates

Beryllium fluoride, BeF_2, melts at the comparatively low temperature of 540°C to a highly viscous liquid. Mackenzie[66] measured its viscosity and electric conductivity over the range 700–950°C, and found that the activation energy for viscous flow increased from 160 KJ mol^{-1} at 920° to 290 KJ mol^{-1} at 700°C. The viscosity near the melting temperature is greater than 10^5 N s m^{-2}, and its electrical conductivity is only 10^{-8} S cm^{-1}, both values being typical of a high molecular weight, covalently bonded polymer; but Mackenzie points out that the low electrical conductivity is not inconsistent with an ionic Be—F bond, since electrical conduction in such a melt may involve the transport of very large units. The activation energy for electrical conduction is similar to that for viscous flow, indicating that both processes involve the movement of similar-sized structures.

When molten beryllium fluoride is cooled, it readily forms a glass(67) which has been shown by X-ray diffraction(68) to contain the same structural units as the known crystalline forms of beryllium fluoride, but in a disordered array. This glass must therefore be a four-connected network polymer composed of tetrahedral BeF_4 units sharing fluorine atoms at their vertices, and is structurally analogous to vitreous silica. The crystalline forms of beryllium fluoride also resemble silica in exhibiting polymorphism, and both a cristobalite form and a quartz form have been identified.(69) Goldschmidt observed(70) that the structural analogy between SiO_2 and BeF_2 extended to their compounds, the silicates, M_2SiO_3, and the fluoroberyllates, $MBeF_3$. Vogel and Gerth(71) prepared several different types of binary fluoroberyllate glasses by melting together alkali or alkaline earth fluorides with beryllium fluoride, and determined the range of glass-forming compositions for most of these systems. They also found that many of these apparently simple, homogeneous glasses exhibited clearly marked phase separation in the electron microscope; for example glasses of the lithium fluoride-beryllium fluoride type with 15 mol % LiF separated into two interconnected phases. Above 25 mol % LiF, isolated droplets of the disperse phase were formed which decreased in both size and quantity with increasing lithium fluoride content. The disperse phase crystallized more readily than the matrix.

More complex fluoroberyllate glasses have also been prepared, and some optical glasses containing aluminium fluoride, lead fluoride and beryllium fluoride have been described.(72)

Heyne(73) prepared a number of multicomponent fluoroberyllate glasses containing alkali, alkaline earth metals and aluminium by dissolving mixtures of the appropriate carbonates in hydrofluoric acid and evaporating to dryness, then fusing the residue. All these materials are hygroscopic and decomposed by water. Their infra-red absorption spectra normally contain one intense, broad band centred on 3 μm. Schröder(74) has made a series of fluoride glasses based on hydrogen fluoride, which are remarkable for their stability towards water, but it is by no means certain that these materials should be classified as inorganic polymers although most of them contain potential network-formers such as aluminium fluoride and lanthanum fluoride. Grebenshchikov(75) has reviewed the structure and chemistry of beryllium fluoride and the formation of glassy fluoroberyllate polymers, comparing their properties with those of the analogous silicates. He concluded that although fluoroberyllates and silicates are crystallographic analogues, there are significant differences in the nature of the glasses.

Phosphorus Oxynitride

Phosphorus oxynitride $(PON)_x$, was first prepared by Gerhardt in 1846[76] by heating phosphoryl imidoamide, a partial condensation product of phosphorylamide, at red heat. It is obtained as a white powder by heating phosphorylamide at temperatures of 600°C or higher; above 1000°C, pure phosphorus oxynitride melts to a transparent glass. It has also been made by heating phosphorus pentachloride with ammonium chloride in the presence of oxygen at 925–950°C and then in ammonia at 750–800°C.[77] Because the products from these condensations are all amorphous to X-rays and the structural information that has been obtained by infra-red spectroscopy is incomplete, its detailed structure is not known with certainty. Steger and Mildner[78] studied the process of condensation of phosphorylamide to phosphorus oxynitride with the aid of infra-red spectroscopy and reached the conclusion that the polymer probably contains P—O—P and P—N—P linkages. The fact that polydichlorophosphazene reacts at elevated temperatures with metal oxides to give an almost identical material suggests that phosphorus oxynitride probably contains —P=N—P— chains cross-linked through oxygen, and its similarity to silica in chemical inertness, high melting point and glass-forming ability is a strong temptation to propose the four-connective network structure

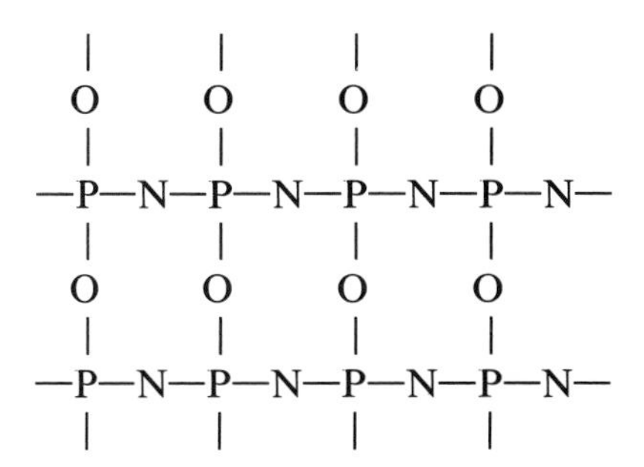

which is isoelectronic with silica,[79] but on present evidence other possibilities cannot be ruled out.

Phosphorus oxynitride can be drawn into fibres from the melt, and will react with metal oxides to give glasses of somewhat lower softening point and improved resistance to hydrolysis. It has been proposed as a binder for asbestos fibres[80] to form ceramic-like materials by moulding a mixture of phosphorylamide and chrysotile fibres under pressure at 460–470°C, then heating the product at 650–875°C until a weight loss corresponding to complete condensation to $(PON)_x$ has occurred. Phosphorus oxynitride is not attacked by aqueous acids or alkalies, but is decomposed by fusion with caustic soda to give ammonia and sodium phosphate, and by concentrated sulphuric acid to give ammonium phosphate. It is reported[79] that it decomposes slowly in a vacuum at 750°C, although it appears to be stable at this temperature in air at atmospheric pressure.

CRYSTALLINE FOUR-CONNECTIVE NETWORK POLYMERS

Crystalline Silicates

The study of crystalline silicates has long been regarded as the special province of geologists and mineralogists, and has received scant attention from polymer scientists; yet this class of compounds is the most abundant of all naturally occurring polymers, greatly exceeding in quantity the total of all organic polymers, both natural and synthetic, on the earth. Crystalline silicates also provide the basis of all our common building materials including bricks, cement, concrete, slates and tiles, as well as most of our common household utensils such as cups, saucers and plates.

All silicates are built up from the same four-connective SiO_4 unit and can therefore be classified as four-connected network polymers, although the possible arrangements of SiO_4 units in different crystalline silicates are as varied as the possible arrangements of carbon atoms in organic compounds. There are crystalline silicates with structures that are analogous to linear hydrocarbon chains, fused-ring aromatic polymers and graphite. The crystalline silicates are usually classified according to the complexity of the anions present in their crystal lattice. Apart from the orthosilicates, such as olivine and garnet (which contain isolated SiO_4^{4-} units) and the linear and cyclic oligomers, such as thortveitite (a simple disilicate) and beryl (which contains a six-membered ring anion $Si_6O_{18}^{12-}$), the vast majority are ionic polymers of virtually infinite molecular weight, associated with a variety of different cations. From the viewpoint of polymer science they can be classified into linear chain polymers, ladder polymers, sheet or layer polymers and three-dimensional networks.

Linear Chain Polymers: Pyroxenes

Chains of linked SiO_4 tetrahedra can be arranged in a number of possible ways, some of which are shown schematically in Fig. 5.7. These are characterized by

Fig. 5.7. Linear chains of SiO_4 tetrahedra with repeat units containing one, two, three, five and seven tetrahedra.

the number of tetrahedra in the repeat length. There is no naturally occurring crystalline silicate with a single tetrahedron repeat unit; but chains with two-, three-, four-, five- and seven-tetrahedra repeat units have all been identified. The largest group of chain silicates are the pyroxenes; these contain chains with a repeat distance of approximately 0·53 nm, which corresponds to two SiO_4 units. The majority of the pyroxenes can be regarded as copolymers of magnesium, calcium and iron metasilicates with empirical formulae of the type ($Mg_x Ca_y Fe_{1-x-y}$)SiO_3. In addition there are a number of important pyroxenes that contain alkali cations, such as aegirine $NaFe^{(+3)}Si_2O_6$, spodumene $LiAlSi_2O_6$ and jadeite $NaAlSi_2O_6$. The structure of a typical pyroxene is shown in Fig. 5.8. The chains are linked laterally by cations but can have various dispositions relative to one another, leading to sub-divisions of the pyroxenes into clino- and orthopyroxenes. The first pyroxene to have its structure completely determined was diopside.[81] In diopside the pyroxene chains are linked laterally by calcium and magnesium ions arranged in such a way that the magnesium cations are octahedrally co-ordinated to oxygen atoms that are linked to only one silicon atom, while the larger calcium ions are surrounded by eight oxygen atoms, two of which are shared by adjacent SiO_4 tetrahedra. Many diopsides contain appreciable amounts of chromium, and a nickel diopside, $CaNiSi_2O_6$, has been synthesized.[82] The melting point of anhydrous diopside is 1391°C,[83] but under a pressure of 5 kbars of water vapour it is lowered to 1290°C; the latent heat of fusion is 422 $J g^{-1}$, and its specific gravity is 3·22 to 3·38.

Spodumene, $LiAlSi_2O_6$, is another pyroxene that has been fully characterized. This mineral is occasionally found in the form of giant crystals; single crystal specimens up to 2 metres thick and 12 metres long have been found in South Dakota. The clear, emerald-green variety containing a chromium impurity and known as hiddenite is regarded as a gemstone. Its structure is similar to that of diopside, and like diopside it can be produced synthetically—for example by heating gels of the composition Li_2O, Al_2O_3, 4–8 SiO_2 at 360°C for several weeks in an autoclave.

Wollastonite differs from the pyroxenes in having a repeat distance along the silicate chain of 0·73 nm, coresponding to three SiO_4 tetrahedra. This is a common mineral found in metamorphosed limestones, which has the ideal formula $CaSiO_3$, although it often also contains iron and manganese, together with smaller amounts of magnesium. Wollastonite can be produced by the reaction of quartz with calcite in the temperature range 600–800°C, and is readily synthesized from hydrous gels of the right composition to form xonotlite, $Ca_6(Si_6O_{17})(OH)_2$, which readily decomposes on heating to give wolastonite.

Silicate chains with a repeat distance of five tetrahedral units are found in the mineral rhodonite ($MnSiO_3$, normally contaminated with calcium) which is a

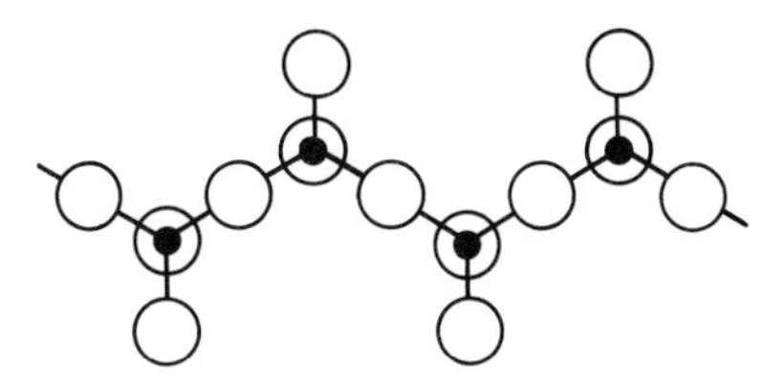

Fig. 5.8. Idealized pyroxene structure.

pink or reddish mineral found in association with manganese-bearing ore bodies in Australia, New Zealand, Japan, Sweden and the United States. A fine-grained rose-pink variety from New South Wales has often been cut and polished for use as an ornamental stone. Another manganese silicate called pyroxmangite, which has a similar composition to rhodonite, contains a silicate chain with repeat units of seven tetrahedra, the repeat distance being 1·74 nm. Pyroxmangite often contains small amounts of iron, calcium and magnesium as impurities. This mineral is usually pale pink or lilac in colour, although specimens as found may be dark brown to black due to a film of altered material on the surface.

Ladder Polymers: Amphiboles

The crystalline silicates which contain a polymeric anion built up of two linked chains of SiO_4 tetrahedra are called amphiboles (Fig. 5.9). These double chains have a repeat distance along their length of approximately 0·53 nm and this defines one dimension of the unit cell. They are separated from and bonded to each other laterally by planes of cations, but in addition there may be hydroxyl groups present in the structure. The sizes of the cations determine the way in which they are co-ordinated to the oxygen atoms in the chains and this determines the relative positions of adjacent "ladders" in the crystal. Generally the stacking of the double chains is such as to produce a monoclinic unit cell containing the $Si_8O_{22}^{16-}$ anion. In amphiboles containing only small cations,

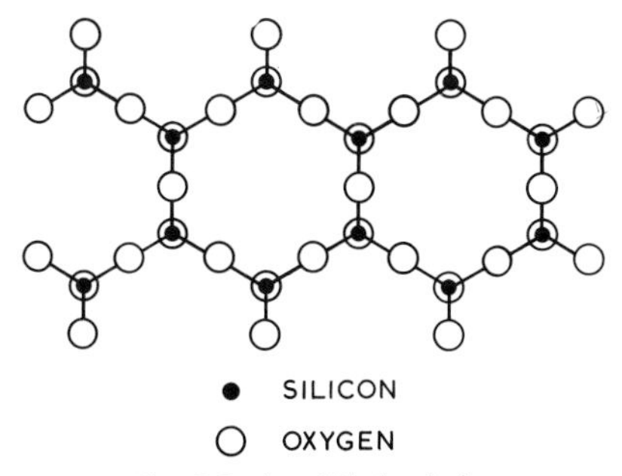

Fig. 5.9. Amphibole chain.

such as magnesium, there is instead another more favourable stacking that produces an orthorhombic unit cell.

The amphibole structure allows great variety of chemical composition as a result of ionic replacement, and not only is there considerable interchange of cations between different amphiboles, but it is also common to find silicon replaced by aluminium (although this substitution appears to be limited to two aluminium atoms per formula unit), and hydroxyl groups may be partially or wholly replaced by fluorine and chlorine. There are, therefore, very many different amphibole minerals, and their classification is a complex matter outside the scope of this book. It will suffice for the present purpose to discuss the structure and properties of four typical amphiboles that are of special interest to polymer scientists and technologists because of their fibrous nature; these are anthophyllite, amosite, crocidolite and tremolite which, together with chrysotile (a two-dimensional sheet polymer to be discussed in the following section), make up the group of minerals that is known collectively as asbestos.

Anthophyllite has the ideal composition $(Mg, Fe)_7Si_8O_{22}(OH)_2$ and is an orthorhombic amphibole that is found as fibrous deposits in the United States, Canada, Sweden and Finland. A pure magnesium anthophyllite ($Mg_7Si_8O_{22}(OH)_2$) has been synthesized[84] by heating talc for an hour at 800°C under a water -vapour presure of 1000 bars, but on continued heating the anthophyllite decomposed to enstatite (a pyroxene) and quartz, showing that the orthorhombic amphibole is unstable under these conditions. Boyd[85] has reported the synthesis of a fibrous anthophyllite containing 40–50% of $Mg_7Si_8O_{22}(OH)_2$. Anthophyllite contracts reversibly by about 0·45% at 830°C; its fibres are rather weak and consequently it is of less industrial importance than the other asbestos minerals.

Amosite and montasite (which is the same mineral) are monoclinic amphiboles containing a much higher proportion of iron than anthophyllite. They are mainly found in South Africa (the name amosite is derived from the initials of Asbestos Mines of South Africa), also in Finland, New Zealand, the U.S.S.R. and Scotland. The fibres are usually of rather a harsh texture, but the montasite variety (from the Montana Mine in Transvaal) is soft and silky. Amosite asbestos typically has a density of 3·1 $g\,cm^{-3}$, a tensile strength of around 2 $GN\,m^{-2}$ and a Young's modulus of 150 $GN\,m^{-2}$. The individual fibres are from 60–100 nm wide.

Crocidolite is the most important of the amphibole varieties of asbestos. It has the ideal composition $Na_2Fe_3^{II}\ Fe_2^{III}\ Si_8O_{22}(OH)_2$ and, like amosite, is monoclinic. This mineral is widely distributed; it is found in South Africa, Western Australia, Rhodesia, Wyoming, Colorado and Utah, as well as the U.S.S.R. It is usually blue or indigo in colour and for this reason it is commonly called blue asbestos; it exhibits marked pleochroism, the absorption colours being dark blue parallel to the fibre axis, blue-grey to yellow in one direction at right angles to the first and yellowish green in the third direction. The density of pure crocidolite is $3{\cdot}42\ g\,cm^{-3}$, its tensile strength is around $3{\cdot}5\ GN\,m^{-2}$ and its Young's modulus is $175\ GN\,m^{-2}$.

Both crocidolite and amosite are nearly neutral in reaction and resistant to acids, in contrast to chrysotile asbestos (see below) which is strongly basic in reaction and decomposed by acids.

Tremolite has the idealized composition $Ca_2Mg_5Si_8O_{22}(OH)_2$. This is another monoclinic, fibrous amphibole that is readily distinguished from crocidolite by its lack of colour and pleochroism. Tremolite is found in the Balkans, Sierra Leone, Italy, Japan, the United States, Devon and Scotland; it is an early product of metamorphism of dolomitic limestones containing silica impurities, the other reaction product being calcite. Like anthophyllite, tremolite contracts by about 0·65% in the fibre direction on heating at 825°C, but in the case of tremolite the contraction is irreversible.

All the amphibole fibres lose their structurally combined water on heating above 400°C and the fibres are embrittled. Their decomposition products melt in the range 1200–1400°C.

Two-dimensional Sheet Polymers: the "Layer" Silicates

The layer silicates or phyllosilicates are built up of composite layers consisting of sheets of the quasi-infinite two-dimensional anion $(Si_4O_{10}{}^{4-})_n$ bound either to one or to both sides of a single sheet of the layer hydroxides gibbsite, $Al(OH)_3$, or brucite, $Mg(OH)_2$. The magnesium or aluminium ions are octahedrally co-ordinated to oxygen atoms in the silicate network. Any residual charge on the composite sheet is satisfied by additional cations, typically alkali metal, alkaline earth, or iron cations, located between the layers; these cations are often readily exchangeable under quite mild conditions and the spacing between the layers varies with the size and the state of hydration of the cations.

It is useful to classify the layer silicates into single-sided structures and double-sided or "sandwich" structures. Single-sided structures are those which are built up of just one silicate sheet bound to each layer of brucite or gibbsite, whereas the double-sided structures consist of a layer of brucite or gibbsite sandwiched between two silicate sheets (Fig. 5.10). A further classification can be made according to the nature of the non-silicate component, that is whether it is derived from brucite $[Mg(OH)_2]$ or gibbsite $[Al(OH)_3]$. The relationship between some of the more important layer silicates is shown by this method of classification in Table XV.

Single-sided Layer Structures

Chrysotile. This is the most important member of the serpentine group of

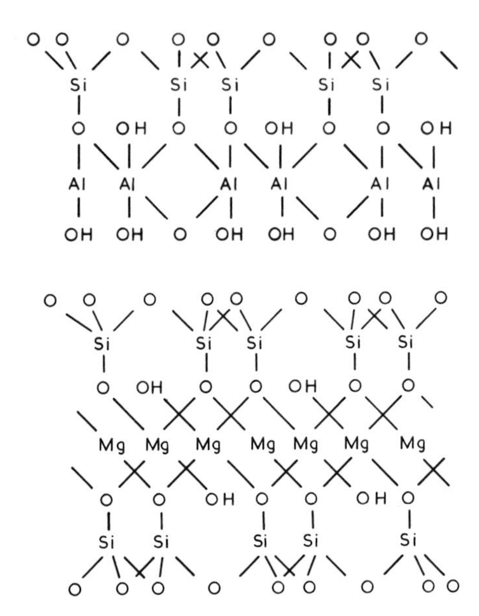

Fig. 10. Single-sided (kaolinite) and double-sided (vermiculite) layer silicate structures.

Table XV
Classification of layer silicates

Single-sided layers	Sandwich layers	Second component
Chrysotile	Vermiculite, Talc	Brucite, $Mg(OH)_2$
Kaolinite	Montmorillonite	Gibbsite, $Al(OH)_3$
Halloysite	Pyrophyllite	

minerals. It has the approximate composition $Mg_3Si_2O_5(OH)_4$ and consists of a silicate sheet made up of a hexagonal network of SiO_4 tetradedra linked in two dimensions, joined to a brucite sheet in which two out of every three hydroxyl groups are replaced by oxygen atoms from the SiO_4 tetrahedra (Fig. 5.11). The distance between the sheets is about 0.73 nm. The repeat distances in a planar silicate sheet (0·50 and 0·87 nm) are significantly smaller than those in brucite (0·54 and 0·93 nm) so that there is a considerable mismatch between the two components, and in chrysotile this mismatch is accommodated by curvature of the composite layer with the brucite component that has the larger repeat distances on the outside of the curve. This results in a tubular morphology, and chrysotile fibres can be seen in an electron microscope to be bundles of hollow tubes. The individual fibrils may have various diameters, but typically they have

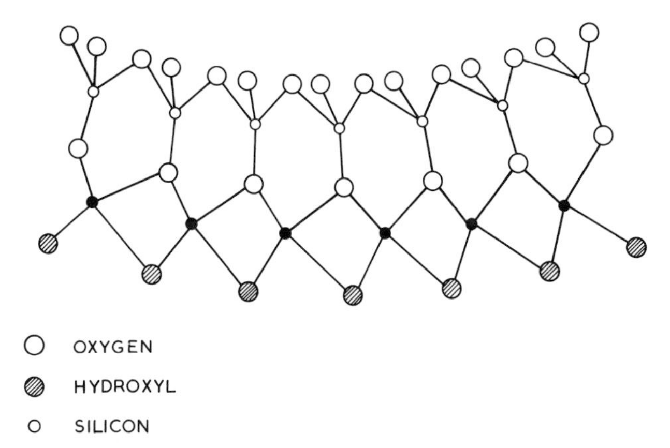

Fig. 5.11. Structure of chrysotile.

an inner diameter of 11·0 nm, and an outer diameter of 26 nm, with ten layers in the wall of the tube. This gives an average radius of curvature of 7·5 nm which may be compared to the calculated value of 8·8 nm for a single strain-free, composite layer.

Chrysotile is the most important source of commercial asbestos; it occurs in veins of silky fibres, which are usually aligned across the vein, and which can be from 1 to 15 cm in length. Chrysotile is found in Canada, the United States, South Africa, Rhodesia and the U.S.S.R. It accounts for 90% of the world market in asbestos and its major uses are as a reinforcing fibre in combination with cement to make sheet, pipes and other shapes; in combination with PVC to make floor and ceiling tiles; and in combination with phenol-formaldehyde and other resins to make brake linings. The uses of asbestos are, however, likely to decline in the future, especially in inhabited buildings, as a result of the now clearly recognized toxic hazards that are associated with exposure to its dust. It is nowadays generally accepted that this hazard applies to all types of asbestos and not, as was once thought, only to crocidolite or "blue asbestos". Chrysotile from the majority of natural deposits has a measured surface area of only 10–20 $m^2 g^{-1}$, indicating that the pores of the fibres are blocked, presumably by adsorbed water; but Californian varieties of chrysotile can have surface areas of 50–55 $m^2 g^{-1}$. It is readily decomposed by acids and begins to lose chemically bound water in the form of hydroxyl groups at 450–475°C. Above 500°C, the structure collapses, the fibres lose all their strength, and the mineral decomposes into forsterite (Mg_2SiO_4) and amorphous silica. In contrast to the amphibole asbestos minerals such as crocidolite, chrysotile is strongly basic; an aqueous suspension has a pH of 9–10, and for this reason it cannot be used as a reinforcement in those organic plastics that are sensitive to alkaline degradation, such as polyesters.

Tubular chrysotile fibres have been synthesized by hydrothermal reaction of a mixture of oxides with the composition 3 MgO·2 SiO_2 at 400–450°C; substitution of magnesium by other cations, such as nickel for example, reduces the tendency to produce tubules while addition of sodium chloride enhances the tendency towards tubular morphology.

The properties of the four most important fibrous silicates that are classified as asbestos minerals are listed in Table XVI.

Table XVI
Properties of asbestos fibres

Fibre	Density g cm^{-3}	Tensile Strength GN m^{-2}	Young's Modulus GN m^{-2}
Chrysotile	2·58	0·7–2·8	160
Amosite	3·1	2	150
Crocidolite	3·42	3·5	175
Tremolite	2·9–3·2	0·007–0·06	140

Kaolinite and halloysite. Kaolinite is the most important member of a number of closely related minerals which are usually classified together as the kaolinite group. The natural aggregate containing these minerals is known as china clay or kaolin. Other members of the group are dickite, nacrite, anauxite and halloysite. Halloysite frequently occurs in a tubular morphology like chrysotile, but in this case the silicate sheet lies on the outside of the curved layer since it has a larger repeat distance than that in the gibbsite sheet (0·86 nm).

The ideal composition of kaolinite is $Al_4Si_4O_{10}(OH)_8$, and the structure is that of an extended sheet of silicate tetrahedra linked into hexagonal rings, joined on one side to a gibbsite sheet in which one out of every three hydroxyl groups is replaced by an oxygen atom from the SiO_4 tetrahedra. Successive layers are superimposed in such a way that the oxygen atoms at the base of one layer are paired by close approach to the hydroxyl groups at the top of the next one. By reason of these hydrogen bonds between successive layers, the interlayer forces in kaolinite, dickite and nacrite are strong enough to resist the tendency of the individual layers to roll up. In halloysite however, there is a layer of water molecules between the mineral layers which increases the distance between successive layers from the value of 0·72 nm that is typical of kaolinite to about

1·0 nm, and this is sufficient to weaken the interlayer forces sufficiently for curling of the layers to occur, as revealed by electron microscopy.[86]

With the exception of halloysite, the individual members of the kaolinite group of minerals show only a small cation exchange capacity (about 10 mEq 100 g^{-1}). Halloysite has an average ion exchange capacity that is about four times greater, and in addition can intercalate neutral organic molecules which replace the interlayer water molecules.

When the kaolinite minerals are heated to 650°C, all the water molecules and hydroxyl groups are removed and their structure becomes highly disordered. Above 800°C, the structure of the layers is completely disrupted, and at still higher temperatures the minerals decompose, forming mullite, $3\ Al_2O_3 \cdot 2\ SiO_2$ and cristobalite. Kaolinite has been synthesized by several workers from aqueous gels of suitable composition by heating between 250–400°C.[87] The plastic properties of kaolinite and related minerals when mixed with water are well known. There is an optimum concentration of water that produces maximum plasticity, and there is some evidence that in this composition range the water molecules are arranged in a definite pattern, and not in a disordered liquid state; the adsorbed water behaves as if it were both denser and more viscous than ordinary water.

Sandwich Structures

Talc and pyrophillite. Talc is the major constituent of rocks known as steatite or soapstone, which are used as electrical insulators. Its approximate formula is $Mg_6(Si_4O_{10})_2(OH)_4$ and its structure consists of a sheet of brucite, $Mg(OH)_2$, sandwiched between two sheets made up of condensed rings of SiO_4 tetrahedra; two out of every three hydroxyl groups in the brucite sheet are replaced by oxygen atoms from the silicate tetrahedra. Talc has no exchangeable cations and no water molecules intercalated between the layers; it is unaffected by acids and organic liquids, and stable to 700°C. At higher temperatures the layers begin to dissociate, first into amphibole chains, and then into pyroxenes. Finally at 1250°C, clinoenstatite and cristobalite are formed. Talc can be synthesized at temperatures below 800°C by heating equimolar mixtures of magnesium oxide and silica under 400–2000 bars pressure of water vapour.[88]

Talc is found in Norway, Sweden, Italy, the U.S.S.R., India and the United States; it usually occurs in massive aggregates, and large single crystals are rare but have been reported to be flexible and slightly elastic.[89]

Pyrophillite, ideally $Al_4(Si_4O_{10})_2(OH_4)$, has a similar structure to talc, except that in place of a brucite sheet it has a sheet of gibbsite sandwiched between two silicate sheets. The arrangement of the sheets in the sandwich is such that two-thirds of the available octahedral sites are occupied by aluminium and one-third are empty (Fig. 5.12). The resulting layers are electrically neutral, consequently the forces between adjacent layers are weak and the layers are easily displaced. Pyrophillite is therefore a soft mineral like talc; a compact variety known as steatite was once used for slate pencils, and is nowadays used to make electrical insulators.

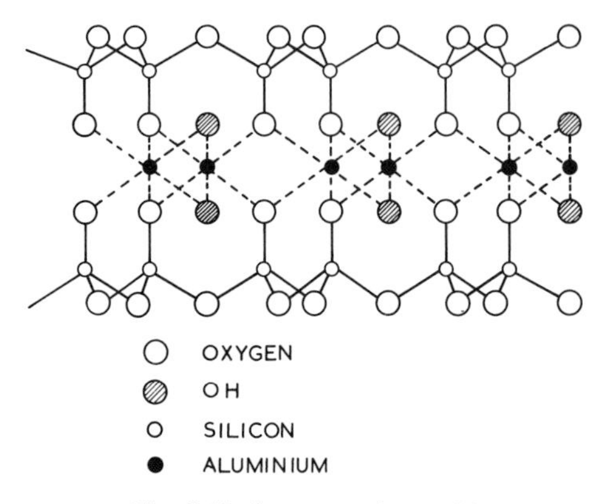

Fig. 5.12. Structure of pyrophillite.

On heating, pyrophillite loses its bound water at 700°C, and at around 1000°C it decomposes into mullite and cristobalite. Pyrophillite is readily synthesized, for example by heating mixtures of hydrated alumina and silica to 400°C. Roy and Osborn[90] showed that pyrophillite is the stable phase in the system Al_2O_3—SiO_2—H_2O over the temperature range 420–575°C, and it may also be stable below 420°C if there is insufficient water present for the formation of higher hydrates. Natural pyrophillite is comparatively rare; it occurs in Japan, Sweden, South Korea and California, and is found in three forms: fine-grained foliated lamellae, radiating granular crystals and needles and compact spherulitic aggregates of small crystals.

Vermiculite and montmorillonite. The structure of vermiculite is similar to that of talc, in so far as the layers consist of brucite sheets sandwiched between two silicate sheets. Whereas in talc, however, the layers are electrically neutral and the cohesion between adjacent layers is weak, in vermiculite there is partial replacement of silicon atoms in the silicate sheets by aluminium, which produces a net negative charge that is balanced by interlayer cations (chiefly magnesium), so that adjacent layers are bound together by much stronger ionic forces. In addition to the cations, there are water molecules between the layers, which form a distorted hexagonal pattern such that each water molecule is linked by a hydrogen bond to an oxygen atom in the silicate sheet. Weak hydrogen bonds also link the water molecules together in the direction of the layers. The basal spacing is normally 1·44 nm, but this can alter both as a result of dehydration and when the interlayer cations are exchanged for others of different size or charge. Dehydration reduces the basal spacing in a series of discrete steps to 0·9 nm, while replacement of the interlayer magnesium ions by, for example, calcium increases it to 1·5 nm. Replacement by potassium or ammonium, on the other hand, reduces the basal spacing to 1·0 and 1·1 nm respectively although these ions are larger, because they carry no hydration shell.

The cation exchange capacity of vermiculite is the largest of all the clay minerals, amounting to between 1·0–2·6 mEq g^{-1}, and vermiculite is the mineral that is mainly responsible for the fixation of potassium in soils. Vermiculite also has the ability to intercalate neutral organic molecules which replace the interlayer water molecules. Vermiculites with magnesium, calcium, sodium and lithium cations can absorb ethylene glycol, but ammonium and potassium vermiculites which have no interlayer water molecules cannot.

Dehydration of vermiculite by heating results in exfoliation; the rapid generation of steam that occurs when the mineral is heated suddenly to about 300°C causes buckling and separation of the layers which can result in a volume expansion of as much as thirtyfold. Exfoliation can also be brought about by treatment with hydrogen peroxide (probably leading to evolution of oxygen as a result of reaction with ferrous ions frequently present as an impurity) and by ion-exchange with organic cations containing long-chain alkyl groups, followed by dispersion in de-ionized water.

Vermiculite is found in South Africa, the United States, Finland and Bavaria; it has not yet been synthesized, but it can be prepared by treatment of phlogopite (a mica) with magnesium chloride solution.

Montmorillonite has basically the same structure as pyrophillite, namely a sheet of gibbsite sandwiched between two silicate sheets, but it differs from pyrophillite in the same way that vermiculite differs from talc. Substitution either of silicon atoms by aluminium atoms in the silicate sheets, or of aluminium atoms by magnesium atoms in the gibbsite sheet leaves the composite layers with a net negative charge, which is balanced by interlayer cations, usually sodium or calcium. Water is also adsorbed between the layers, and the number of layers of water molecules present depends on the nature of the interlayer cations; calcium montmorillonite has two layers of water molecules between adjacent silicate sheets, and sodium montmorillonite may have one, two, three or even more layers. There is, therefore, a continuously variable basal spacing from about 1·0 to about 2·0 nm. Montmorillonite and other similar clay minerals will also absorb organic cations and a variety of organic liquids; substituted ammonium ions and even proteins are taken up by clays giving interlayer spacings ranging from 1·2 to 4·8 nm,[91] and polar organic molecules such as glycol and glycerol are absorbed in integral numbers of monolayers to give characteristic basal spacings.[92]

On heating at temperatures between 100°C and 200°C, montmorillonite loses the interlayer water molecules reversibly, with a decrease in the basal spacing from 1·2 to 1·0 nm. At higher temperatures, chemically bound water in the form of hydroxyl groups in the gibbsite sheet is lost irreversibly and the structure begins to alter drastically.

Montmorillonite is widely distributed throughout Europe and America; it is the main constituent of "Fuller's earth", a clay that has a particularly high absorptive capacity. It has also been synthesized by boiling magnesium chloride and hydrated silica with lime or caustic soda,[93] by treatment of illite (a mica) with magnesium chloride,[94] and by hydrothermal reactions of magnesium oxide, alumina and silica.[95]

Three-dimensional Network Polymers

The three most important types of crystalline polysilicates with a fully developed network structure extending in three dimensions are silica itself, the feldspars and the zeolites.

Silica

Silica crystallizes in three main forms—quartz, tridymite and cristobalite; in addition, three other crystalline varieties have been synthesized at high pressures. The physical properties and stability ranges of the known crystalline

Table XVII
Crystalline forms of silica

Modification	Density	Stability range	Melting point °C
α-Quartz	2·65	up to 573°C	1610
β-Quartz		573°–870C	
δ-Quartz		formed below −183°C	
α-Tridymite	2·27	can exist up to 117°C	
β_1-Tridymite		can exist 117°–163°C	
β_2-Tridymite		870–1470°C	1670
α-Cristobalite	2·32	can exist up to 200°C	
β-Cristobalite		1470–1713°C	1713
Keatite	2·50	formed at 380–585°C and 35–120 $MN m^{-2}$	
Coesite	3·01	formed at 500–800°C and 3·5 $GN m^{-2}$	
Stishovite	4.3	synthesized at >1200°C and 16 $GN m^{-2}$	

forms of silica are given in Table XVII. After the feldspars, quartz is the most abundant mineral in the earth's crust, being both a typical constituent of igneous rocks and at the same time one of the commonest sedimentary deposits in the from of sand and sandstones.

The compacted microcrystalline varieties of silica such as chert, flint, agate and jasper, which are collectively known as chalcedony, are also forms of quartz, being made up of an interlocking network of microcrystals with a multitude of fine pores penetrating the solid. The structure of quartz (Fig. 5.13) is a three-dimensional network of six-membered Si—O rings (i.e. three SiO_4 tetrahedra) linked together in such a way that every six such rings enclose a twelve-

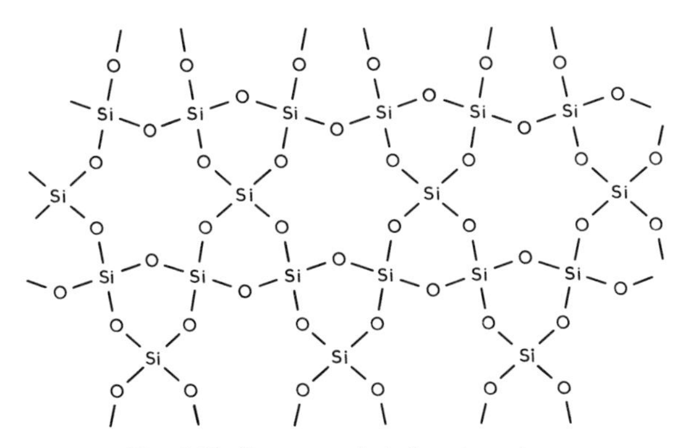

Fig. 5.13. Quartz, projected on to a plane.

membered Si—O ring; a projection of the oxygen atoms on to a plane would show a pattern of triangles with their vertices located at the vertices of regular hexagons. Since the silicon atoms do not lie in the same plane but are alternately displaced above and below it, there is a screw axis in the crystal and quartz therefore exists in right- and left-handed forms. Tridymite has a simpler structure (Fig. 5.14) which consists of SiO_4 tetrahedra linked together in a plane to form flat sheets of six-membered rings; the sheets are linked together through the fourth oxygen atoms of the tetrahedra which point alternately above and below the plane of the sheet. Unlike quartz, tridymite has a very open structure containing channels through which comparatively large ions can pass. Cristobalite is also made up of six-membered siloxane rings linked together into a flat sheet, but successive sheets in the crystal are rotated by 60° relative to one another, so that the repeat distance perpendicular to the plane of the sheets is six SiO_4 units, instead of four as in tridymite. Cristobalite has cubic symmetry and may be regarded as analogous to diamond, with silicon atoms occupying the positions of the carbon atoms, and with an oxygen atom situated midway between every adjacent pair of silicon atoms.

Coesite, Keatite and Stishovite are dense crystalline forms of silica that are only formed at high pressures. Coesite differs from the other crystalline forms of

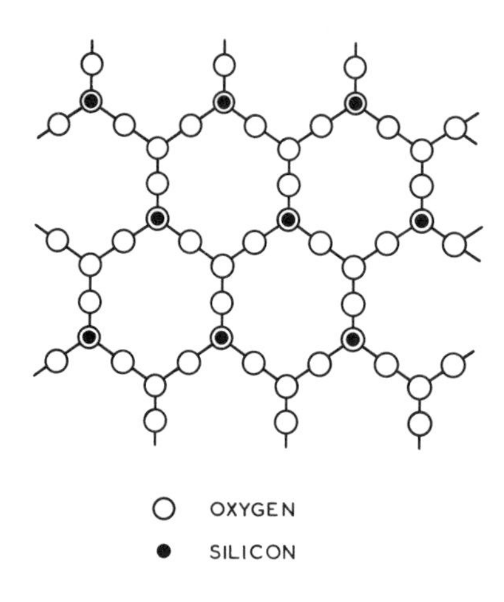

Fig. 5.14. Tridymite.

silica in that it resists attack by hydrofluoric acid; its density (3·01) is considerably higher than that of quartz. It is produced at temperatures in the range 500–800°C and pressures of 35 kbar,[96] and has been found to occur naturally in rocks that have been struck by large meteorites. The network structure of coesite is built up of four-membered rings of SiO_4 tetrahedra, linked three-dimensionally in a somewhat similar manner to the structure of the feldspars (see below).

Keatite is a purely synthetic form of silica produced at 380–585°C and 300–1200 bars;[97] its lattice contains fourfold spirals of SiO_4 tetrahedra which share corners, and the spirals are linked together by additional tetrahedra which bridge four spirals.

Stishovite is formed at pressure of 160 kbar and temperatures above 1200°C; it has also beeen found to occur naturally in Meteor Crater in Arizona.[98]

Feldspars

The feldspars are the most abundant minerals in the earth's crust, accounting for 60% of all igneous rocks, and are second in order of abundance to quartz in sediments. They can be regarded as derivatives of silica in which either one-quarter or one-half of the silicon atoms have been replaced by aluminium, the resulting negative charge being balanced by cations which are either sodium or potassium in the case of the alkali feldspars, and sodium or calcium in the case of the plagioclase feldspars. The majority of feldspars are compounds or solid solutions in the ternary system $Na(AlSi_3O_8)$–$K(AlSi_3O_8)$–$Ca(Al_2Si_2O_8)$, with the polymeric anions $(AlSi_3O_8^-)_n$ and $(Al_2Si_2O_8^{2-})_n$ making up the three-dimensional network structure. The network in a typical feldspar is made up of eight-membered rings, consisting of four $(Al,Si)O_4$ tetrahedra linked by opposite vertices through additional $(Al,Si)O_4$ tetrahedra into chains in two directions (Fig. 5.15). The chains themselves are linked together by rings of four tetrahedra

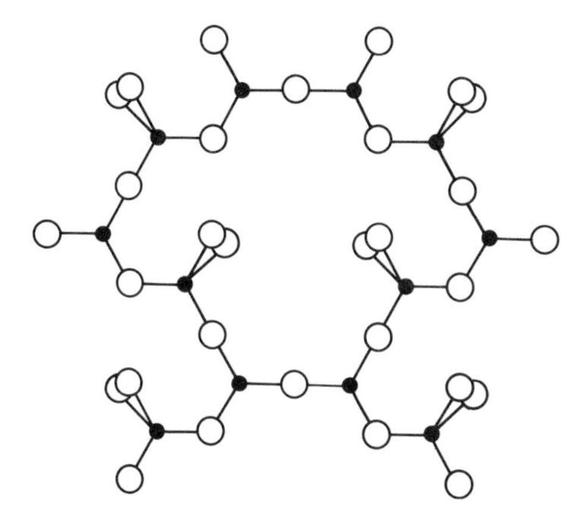

Fig. 5.15. A typical feldspar.

to form a very open framework, the interstices of which are occupied by cations. Since all the Si—O and Al—O distances are the same (0·164 nm), it is probable that the aluminium atoms are randomly distributed. Alkali feldspars have been synthesized by hydrothermal reactions; Wyart[99] prepared orthoclase by reacting muscovite and silica with potassium hydroxide under pressure; Barrer and Hinds synthesized orthoclase from analcite and potassium chloride hydrothermally;[100] Euler and Hellner made a potassium feldspar at 500°C and 1000 bars pressure of water vapour.[101]

Zeolites

The zeolites are also three-dimensional network polymers built up of $(Si,Al)O_4$ tetrahedra linked by sharing of oxygen atoms into rings and cages, and as in the feldspars the overall negative charge of the anion is balanced by sodium, potassium or calcium cations that are located in cavities in the network. The essential difference between the feldspars and the zeolites is in the accessibility of these cavities. The feldspars have relatively compact structures in which the cavities that are occupied by cations are completely surrounded by oxygen atoms, and the cations cannot be moved without breaking Si—O or Al—O network bonds. Furthermore, in feldspars, replacement of a monovalent cation such as Na^+ by a divalent cation such as Ca^{++} requires a change in the Si : Al ratio, and vice versa. The zeolites, on the other hand, have very open structures; the cavities that are occupied by cations are open to the neighbouring cavities on either side, forming interconnecting channels through the network. These channels are large enough for ions and small molecules such as water to pass freely without disrupting the network; moreover, the network is structurally independent of the valency of the cations, since they do not fill all the cavities, and replacement of one divalent cation by two monovalent cations can occur without any change in the network. This characteristic structural feature of the zeolites is displayed in their well-known base-exchange properties, and in their ability to intercalate water and small organic molecules such as the lower paraffins and olefines. Their more open structure is also shown by their much lower densities: the densities of typical zeolites such as mordenite ($Na_2K_2CaAl_2Si_{10}O_{24}\cdot 7H_2O$), chabazite ($CaAl_2Si_4O_{12}\cdot 6H_2O$) and natrolite ($Na_2Al_2Si_3O_{10}\cdot 2H_2O$) lie in the range 2·05–2·20 g cm^{-3}, compared to densities of 2·6–2·7 g cm^{-3} for typical feldspars. Their oxygen packing densities are also characteristically lower than other crystalline silicates; mordenite, for example, contains only 65·1 g atoms oxygen $litre^{-1}$, whereas typical feldspars such as orthoclase and anorthite contain 79–80 g atoms oxygen $litre^{-1}$, and quartz has 88·2 g atoms oxygen $litre^{-1}$.

The zeolites can be subdivided into three main classes; the natrolite group (natrolite, mesolite, scolecite, thomsonite and edingtonite) with a structure built up from rings of four $(Si,Al)O_4$ tetrahedra that are linked together into chains, with substantially fewer linkages between the chains, so that cleavage along the chain direction is preferred and the minerals generally have a fibrous character; the heulandite group (heulandite, stilbite and epistilbite) in which the $(Si,Al)O_4$ tetrahedra form sheets of six-membered rings with relatively fewer linkages between the sheets, so that the minerals tend to exhibit mica-like morphology and cleavage; and thirdly the framework zeolites in which the density of bonding is similar in all three directions—this group includes phillipsite, chabazite, faujasite and many synthetic zeolites.

The fibrous zeolites natrolite, mesolite, scolecite and thomsonite are found in Scotland, Northern Ireland, New South Wales, New Zealand and New Jersey. Thomsonite is frequently encountered in the Isle of Skye. These minerals occur in the form of fine needles, radiating clusters and tufts of fibres. They all contain the polymeric anion $[Al_2Si_3O_{10}{}^{2-}]_n$, which has the structure shown in Fig. 5.16.

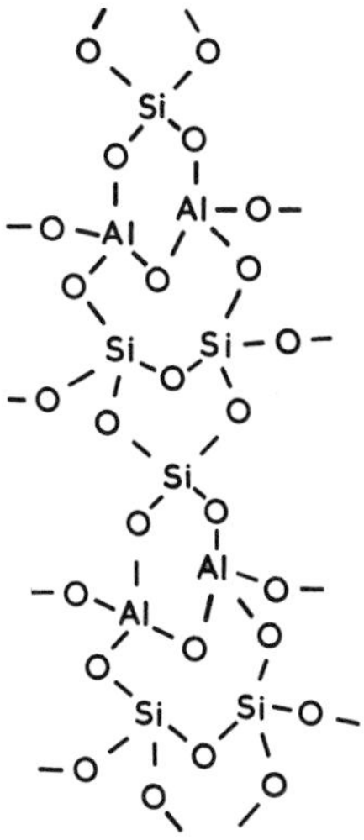

Fig. 5.16. Structure of a typical fibrous zeolite.

Every $(Si,Al)O_4$ tetrahedron shares its oxygen atoms with four others, and the linkages are arranged so that there are chains of alternating six- and eight-membered Si—O—Al rings running through the network, joined laterally through oxygen atoms. Balancing the net negative charge on this polymeric lattice are loosely held alkali and alkaline earth cations, which occupy the channels between the chains. The changes in unit cell dimensions as revealed by X-ray diffraction studies[102] that accompany substitution of smaller cations by larger ones show that the aluminosilicate framework is not completely rigid, but can be distorted (presumably by rotation of units in the chains) to accommodate ions of various sizes.

Certain fibrous zeolites have been synthesized; scolecite $(CaAl_2Si_3O_{10} \cdot 3H_2O)$ is reported to be formed when a mixture of composition $CaO \cdot Al_2O_3 \cdot 3\ SiO_2$ is heated at 230–285°C under 1000 bars of water vapour,[103] but at higher temperatures scolecite is unstable. Thomsonite ($NaCa_2Al_5Si_5O_{10}\cdot 6H_2O$) was synthesized hydrothermally by Goldsmith[104] and by crystallization of a glass of the appropriate composition by Coombs and others.[105]

The sheet-like zeolites heulandite, stilbite and epistilbite are built up of $(Si,Al)O_4$ tetrahedra linked to form a planar network of twelve-membered rings (i.e., rings of six tetrahedra); the unshared oxygen atoms, lying above and below the plane of the sheet, join the sheets together through the remaining silicon and aluminium atoms (Fig. 5.17). This results in a very open structure through which run channels large enough to accommodate large cations such as Cs^+.

Fig. 5.17. Heulandite, a sheet zeolite.

Heulandite and stilbite are commonly found in cavities in basalt, and they have also been synthesized by hydrothermal reactions by Koizumi and Roy;[103] epistilbite has been obtained by crystallization from glasses made from feldspars and excess silica by Coombs and others.[105]

The framework zeolites phillipsite, chabazite, faujasite and mordenite have structures that consist of rings of four, six and eight $(Si,Al)O_4$ tetrahedra joined to form six-sided prisms which enclose large cages (Fig. 5.18). The channels connecting these cavities are of smaller diameter and it is the size restriction that is imposed by the channels that is responsible for the molecular sieve action of these zeolites. For example, in one variety of chabazite, known as levyne, the channels are about 0·27 nm diameter, so that oxygen and nitrogen can be

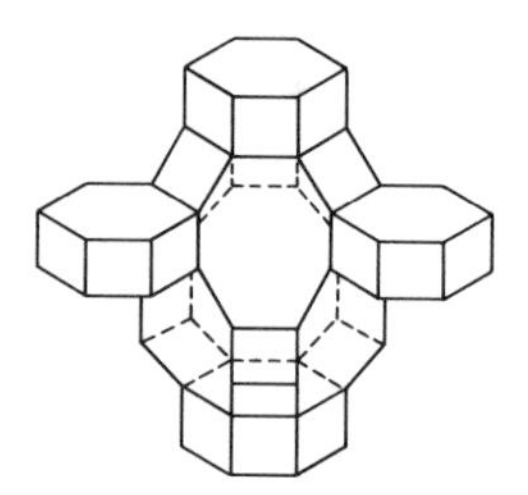

Fig. 5.18. Chabazite, a framework zeolite.

absorbed into the cavities but argon (0·38 nm diameter) is excluded. In chabazite itself the channels are about 0·39 nm wide, and in this case argon, methane and normal paraffins can be absorbed, but branched paraffins and aromatic hydrocarbons are excluded. The structure of faujasite is built up of $(Al,Si)O_4$ tetrahedra linked to form cages similar to those in chabazite, but in this instance the cages are joined by bridges of hexagonal prisms (Fig. 5.19)

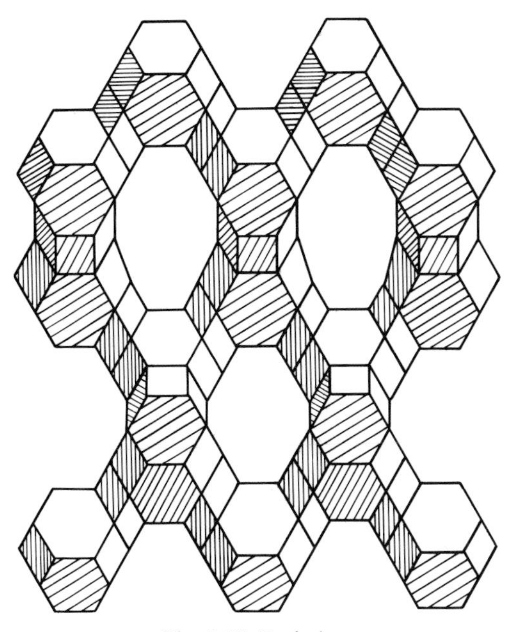

Fig. 5.19. Faujasite.

forming a system of wide channels running throughout the crystal, with a diameter of 0·9 nm. Thus faujasite not only absorbs normal and branched-chain paraffins, but also cyclohexane and the aromatic hydrocarbons such as benzene and toluene.

Many framework zeolites have been synthesized, and synthetic procedures have been extended to the production of zeolites that do not occur in nature. For a detailed account, the reader is referred to the textbook by D. W. Breck.[106] Zeolite synthesis depends upon the use of highly reactive starting materials such as freshly precipitated hydrous gels, a relatively high pH maintained by the presence of caustic alkali, a high degree of supersaturation resulting in large numbers of nuclei, and relatively low temperatures (for example 95°C) to encourage the formation of metastable phases. In a typical procedure, aqueous solutions of sodium aluminate, sodium silicate and sodium hydroxide are mixed in the calculated properties at 25°C and the resulting hydrogel is maintained at a suitable temperature between 25 and 175°C until the desired crystalline phase separates. In spite of the apparent simplicity of this kind of procedure, however, there are many subtleties involved in actual practice, and the reproducible synthesis of any given zeolite often depends critically upon the degree of purity of the materials, their exact proportions, and the presence of suitable seed crystals; some experimenters have even gone to the length of keeping particular glass flasks for the preparation of particular zeolites. As a result of careful synthetic work in this field, the number of different zeolites now known to exist amounts to more than a hundred, and their use extends to many different areas of chemical technology such as: the separation of hydrocarbons; catalytic reactions; dehydration of gases and liquids; carriers for catalysts in plastics and rubber curing; and the removal of both gaseous and liquid pollutants, including sulphur dioxide from the air and radioactive isotopes from solution. As molecular sieves, zeolites are characterized by a uniform pore size ranging from 0·3 to 1·0 nm that is uniquely determined by their crystal structure, a large surface area and a high absorption capacity for gases and vapours. Their surfaces are often very reactive and in the hydrogen form most zeolites are very strong Bronsted acids; they can therefore act as isomerization and cracking catalysts for hydrocarbons.

Boron and Aluminium Phosphates

As a copolymer of the two water-soluble and low melting oxides boric oxide and phosphoric oxide, boron phosphate is remarkable for its insolubility, chemical inertness and high softening point. It is a four-connective crystalline polymer whose structure has been shown by X-ray diffraction to be similar to that of silica, with which it is isoelectronic (Fig. 5.20). The phosphorus-oxygen and boron-oxygen bond lengths are not quite equal, being 0·155 and 0·144 nm respectively, and both the boron and the phosphorus atoms are tetrahedral.

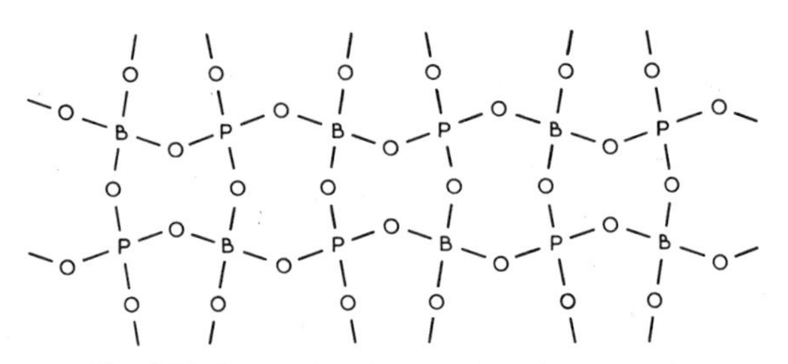

Fig. 5.20. Boron phosphate, projected on to a plane.

Boron phosphate is formed by evaporating mixed solutions of boric and phosphoric acid and heating the residue to 800°C.[107], by reaction of an phosphate with boron trichloride,[108] or by addition of boric oxide to hot phosphoric acid. As first precipitated from aqueous solutions, boron phosphate is soluble in (or decomposed by) water, but after heating above 500°C it becomes insoluble and is then stable to acids and only slowly attacked by concentrated alkalies. Boron phosphate has no true melting point; at 1440–1450°C it sublimes as a mixture of boric and phosphoric oxides. The parent compound BPO_4 does not exist in the glassy state, but glass-forming systems containing BPO_4 and one or more alkali oxide are known (see Chapter 4, p.79, Borophosphate Glasses) and a glass can be prepared from BPO_4—$AlPO_4$—SiO_2 mixtures.[109] Boron phosphate gels in concentrated sulphuric acid are readily obtained, for example by mixing phosphoric acid with a solution of boric oxide in sulphuric acid, and this system has been proposed as a non-spillable electrolyte for lead-acid accumulators.[110]

The Raman spectrum of boron phosphate consists of two main bands at 490 cm^{-1} (BO_4) and 1130 cm^{-1} (PO_4) in the intensity ratio of 2 : 1, together with weaker bands at 240 cm^{-1} (P—O—B) and 1080 cm^{-1} (BO_4). When precipitated at temperatures below 200°C, boron phosphate gives a weak Raman signal at 810 cm^{-1}, indicating the presence of some trivalent boron, but after heating at temperatures above 500°C when it becomes insoluble and resistant to acids, the Raman signal at 810 cm^{-1} disappears showing that all the boron atoms are tetravalent. The infra-red spectrum of boron phosphate contains four strong, rather broad bands centred on 550, 615, 925 and 1085 cm^{-1}.

Aluminium phosphate has the same structure as boron phosphate and in its crystalline modifications it resembles silica even more closely. Quartz-like (Berlinite), tridymite-like and cristobalite-like forms of aluminium phosphate are known and these have somewhat similar stability ranges and transformation temperatures to their corresponding silica analogues.[111] Unlike boron phosphate, aluminium phosphate also melts to a viscous liquid at 1700°C, and the melt readily forms a glass on cooling. In this context it is noteworthy that the averages of the radii and charges of the Al^{3+} and P^{5+} ions are practically identical to the radius and charge of the Si^{4+} ion.

Aluminium phosphate can be prepared by precipitation from aqueous solutions of an alkali or ammonium phosphate and an aluminium salt, such as ammonium aluminium sulphate; the product formed in the range of pH 3–6 is the hydrate $AlPO_4 \cdot 2H_2O$ which on heating at 220°C yields the amorphous, anhydrous compound $AlPO_4$. On further heating at 400°C this crystallizes in the tridymite form, and at about 800°C it is converted into the cristobalite;[112] berlinite, the quartz form, is produced by crystallization at lower temperatures. Amorphous aluminium phosphate is also formed by dissolving freshly precipitated aluminium hydroxide in orthophosphoric acid, by reaction of aluminium alkoxides with phosphoryl chloride or phosphoric acid, and in a curious reaction in which finely divided suspensions of aluminium hydroxide and phosphoric oxide in kerosene are mixed together, when a concentrated phosphoric acid solution of aluminium phosphate separates out as a lower layer.[113]

Amorphous aluminium phosphate is soluble in acids, and forms viscous solutions in phosphoric acid and concentrated aqueous citric acid which are film-forming. It has been used as a binder for abrasives, as a refractory cement, and as a constituent of a variety of catalysts. The crystalline forms of aluminium phosphate, like crystalline boron phosphate, are insoluble in acids, but slowly attacked by concentrated alkalies.

References

1. Rawson, H. (1967). "Inorganic Glass-forming Systems". Academic Press, London.
2. Stanworth, J. E. (1950). "Physical Properties of Glass". Clarendon Press, Oxford.
3. Morey, G. W. (1954). "The Properties of Glass". A. C. S. Monograph no. 124. Reinhold, New York.
4. Pauling, L. (1940). "The Nature of the Chemical Bond". Oxford University Press.
5. Bridgman, P. W. and Simon, I. (1953). *J. Apply. Phys.* **24**, 405.
6. Doremus, R. H. (1962). "Modern Aspects of the Vitreous State", Vol. 2 (Ed. Mackenzie, J. D.) Butterworths, London.
7. Scholze, H. (1969). "VIIIth International Congress on Glass", p. 69. Society of Glass Technology, Sheffield.
8. Zachariasen, W. H. (1932). *J. Amer. Chem. Soc.* **54**, 3841.
9. Warren, B. E. (1938). *J. Amer. Ceram. Soc.* **21**, 259.
10. Konnert, J. H. and Karle, J. (1973). *Acta cryst.* **A29**, 702.
11. Gaskell, P. H. and Howie, A. (1974). In "Proceedings of the 12th International Conference on Physics Semiconductors, 1974", p. 1076. (Ed. Pilkuhn, M. H.) Teubner, Stuttgart.
12. Chaudhari, P., Graczyk, J. F. and Herd, S. R. (1972). *Phys. stat. solidi,* B**51,** 801 and 1973, B**58**, 163.
13. Evans, D. L. and King, S. V. (1966). *Nature,* **212**, 1353.
14. Bell, R. J. and Dean, P. (1972). *Phil. Mag.* **25**, 1381.
15. Da Silva, J. R. G., Pinatti, D. G., Anderson, C. E. and Rudee, M. L. (1975). *Phil. Mag.* **31**, 713.
16. Mozzi, R. L. and Warren, B. E. (1969). *J. Appl. Crystallography,* **2**, 164.
17. Bell, R. J. and Dean, P. (1968). *J. Phys. Chem.* **1,** 299.
18. Konnert, J. H. and Karle, J. (1972). *Acta. cryst.* A**28,** S128.
19. Wright, A. C. and Leadbetter, A. J. (1976). *Phys. Chem. Glasses,* **17**, 122.
20. Vail, J. G. (1952). "Soluble Silicates". A.C.S. Monograph No. 116. Reinhold, New York.
21. Debye, P. and Neumann, R. (1951). *J. Phys. Colloid Chem.* **55**, 1.
22. Weldes, H. H. and Lange, K. R. (1968). "Symposium on glass and related materials", part III. American Society of Chemical Engineers, Philadelphia.
23. Debye, P. and Neumann, R. (1949). *J. Chem. Phys.* **17**, 664.

24. Bacon, L. R. Unpublished Work reported by Vail.[20]
25. Kauzmann, W. (1948). *Chem. Rev.* **43**, 219.
26. Bockris, J. O'M., Mackenzie, J. D. and Kitchener, J. A. (1955). *Trans. Faraday Soc.* **51**, 1734.
27. Mackenzie, J. D. (1956). *Chem. Rev.* **56**, 455.
28. Endell, K. and Hellbrügge, J. (1942). *Naturwissenschaften,* **30**, 42.
29. Douglas, R. W. (1947). *J. Soc. Glass Tech.* **31**, 50.
30. Bockris, J. O'M. and Lowe, D. C. (1954). *Proc. Roy. Soc.* A**226**, 423.
31. Tomlinson, J. W., Heynes, M. S. R. and Bockris, J. O'M. (1958). *Trans. Faraday Soc.* **54**, 1822.
32. Etchepare, J. (1970). *J. Chim. Phys. Physico-Chim. Biol.* **67**, 890.
33. Wilmot, G. B. Ph.D. Thesis, Massachusetts Institute of Technology. Reported (1960). *In* "Modern Aspects of the Vitreous State", (Ed. Mackenzie, J. D.), Vol. 1, Ch. 6. Butterworths, London.
34. Simon, I. and McMahon, H. O. (1953). *J. Amer. Ceram. Soc.* **36**, 160.
35. Porai–Koshits, E. A. (1958). *In* "Proceedings of a Conference on the Structure of Glass, Leningrad Nov. 23–27, 1953". Consultants Bureau, New York.
35. Guaker, R. and Urnes, S. (1973). *Phys. Chem. Glasses,* **14**, 21.
37. Katsuki, K. and Aoki, K. (1970). *Yogyo Kyokai Shi,* **78**, 268.
38. Andreev, N. S. and Porai–Koshits, E. A. (1970). *Discuss. Faraday Soc.* **50**, 135.
39. Neilson, G. F. (1972). *Phys. Chem. Glasses,* **13**, 70.
40. Porai–Koshits, E. A., Goyanov, A. and Averjanov, V. I. (1965). *In* "Physics of Non-crystalline Solids". (Ed. Prins, J. A.) Elsevier, Amsterdam.
41. Hammel, J. J. (1965). Proceedings of the VIIth International Congress on Glass, Brussels, Paper no. 36.
42. Cahn, J. W. (1965). *J. Chem. Phys.* **42**, 93.
43. Seward, T. P., Uhlmann, D. R. and Turnbull, D. (1968). *J. Amer. Ceram. Soc.* **51**, 278.
44. Haller, W., Blackburn, D. H. and Simmons, J. H. (1974). *J. Amer. Ceram. Soc.* **57**, 120.
45. Neilson, G. F. (1972). *Phys. Chem. Glasses,* **13**, 70.
46. Mazurin, P. V., Roskova, G. P. and Kluyev, V. P. (1970). *Discuss. Faraday Soc.* **50**, 191.
47. Redwine, R. W. and Field, M. B. (1968). *J. Materials Sci.* **3**, 380.
48. Uhlmann, D. H. and Kohlbeck, A. G. (1976). *Phys. Chem. Glasses,* **17**, 146.
49. Shaw, R. R. and Uhlmann, D. R. (1968). *J. Amer. Ceram. Soc.* **51**, 377.
50. Jellyman, P. E. and Procter, J. P. (1955). *Trans. Soc. Glass Tech.* **39**, T173.
51. Gross, E. F. and Romanova, M. *Phys.* **2**, 1929, **55**, 744.
52. Langenberg, R. (1937). *Ann. Phys.* **28**, 104.
53. Simon, I. (1960). *In* "Modern Aspects of the Vitreous State", Vol. I, Ch. 6. (Ed. Mackenzie, J. D.) Butterworths, London.
54. Sullivan, E. C. and Taylor, W. C. (1915, 1919). U.S. patents 36 136 and 1 304 623.
55. Flory, P. J. (1953). "Principles of Polymer Chemistry", p. 555. Cornell University Press, New York.
56. Hood, H. P. and Nordberg, M. E. (1934). U.S. patent 2 106 744.
57. Schwertz, F. A. (1949). *J. Amer. Ceram. Soc.* **32**, 390.
58. Saunders, J. B. (1942). *J. Res. Nat. Bur. Stand.* **28**, 51.
59. Rockett, J. J. and Foster, W. R. (1965). *J. Amer. Ceram. Soc.* **48**, 75, 329.
60. Skatulla, W., Vogel, W. and Wessel, H. (1958). *Silikattechnik,* **9**, 51.

61. Bokin, P., Galakhov, F. and Stepanov, E. N. (1971). IXth International Congress on Glass, Versailles, A1.3, p. 335.
62. Zhdanov, S. P. (1959). *Izv. Akad. Nauk. S.S.S.R.* **6**, 1011.
63. Réamur, M. (1793). *Memoires de l'Acad. Sci.* 370.
64. McMillan, P. W. (1964). "Glass-ceramics". Academic Press, London.
65. Maurer, R. D. (1962). *J. Appl. Phys.* **33**, 2132.
66. Mackenzie, J. D. (1960). *J. Chem. Phys.* **32**, 1150.
67. Heyne, G. (1933). *Angew. Chem.* **46**, 473.
68. Warren, B. E. and Hill, C. F. (1934). *Z. Kristallog.*, Abt.A, **89**, 481.
69. Roy, D. M., Roy, R. and Osborn, E. F. (1953). *J. Amer. Ceram. Soc.* **36**, 185.
70. Goldschmidt, V. M. (1926). *Skrifter Norske Videnskaps Akad.* (*Oslo*) *I. Math.-naturwiss. Kl.* **7**, no. 8.
71. Vogel, W. and Gerth, K. (1958). *Glastechn. Ber.* **31**, 15.
72. Sun, K. H. and Callear, T. E. (1949). U.S. patent 2 466 506.
73. Heyne, G. (1933) *Angew. Chem.* **46**, 473.
74. Schröder, J. (1964). *Angew. Chem.* **3**, 376.
75. Grebenshchikov, R. G. (1970). *Arm. Khim. Zh.* **23**, 1068.
76. Gerhardt, C. (1846). *Ann. Chim.* (*Phys.*) **18**, 188.
77. U.S.S.R. patent 254 492 (1969).
78. Steger, E. and Mildner, G. (1964). *Z. anorg. Chem.* **332**, 314.
79. Bloomfield, P. R. (1961). "Thermal Degradation of Polymers", S.C.I. Monograph No. 13, p. 89. Society of Chemical Industry, London.
80. Trainor, J. T. and Kaufman, E. R. (1968) U.S. patent 3 387 980.
81. Warren, B. E. and Bragg, W. L. (1928). *Z. Krist.* **69**, 168.
82. Gjessing, L. (1941). *Norsk. Geol. Tidsskrift,* **20**, 265.
83. Adams, L. H. (1914). *J. Amer. Chem. Soc.* **36**, 65.
84. Bowen, N. L. and Tuttle O. F. (1949). *Bull. Geol. Soc. Amer.* **60**, 439.
85. Boyd, F. R. (1955). *Ann. Rep. Dir. Geophys. Lab. Carnegie Inst.* 117.
86. Bates, T. F. and Comer, J. J. (1959). "Clays and Clay Minerals". Monograph no. 2. Pergamon Press, London.
87. Noll, W. (1944). *Koll. Zeit.* **107**, 81.
88. Noll, W. (1950). *Z. anorg. Chem.* **261**, 1.
89. Deer, W. A., Howie, R. A. and Zussman, J. (1962). "Rock-forming Minerals", Vol. 3, p. 126. Longman, London.
90. Roy, R. and Osborn, E. F. (1954). *Amer. Min.* **39**, 853.
91. Grim. R. E., Allaway, W. H. and Cuthbert, F. L. (1947). *J. Amer. Chem. Soc.* **30**, 137.
92. Bradley, W. F. (1945). *J. Amer. Chem. Soc.* **67**, 975; MacEwan, D. M. C. (1948). *Trans. Faraday Soc.* **44**, 349.
93. Strese, H. and Hofman, U. (1941). *Z. anorg. Chem.* **247**, 65.
94. White, J. L. (1951). *Proc. Soil. Sci. Soc. Amer.* **15**, 129.
95. Roy, R. and Sand, L. B. (1956). *Amer. Min.* **41**, 505.
96. Coes, L., Jr. (1953). *Science,* **118**, 131.
97. Keat, D. D. (1954). *Science,* **120**, 328.
98. Stishov, S. M. and Popova, S. V. (1961). *Geokhimiya,* **10**, 837.
99. Wyart, J. (1947). *Compt. rend.* **225**, 944.
100. Barrer, R. M. and Hinds, L. (1950). *Nature,* **166**, 562.
101. Euler, R. and Hellner, E. (1961). *Z. Krist.* **115**, 433.
102. Hey, M. H. (1932). *Min. Mag.* **23**, 243.
103. Koizumi, M. and Roy, R. (1960). *J. Geol.* **68**, 41.

104. Goldsmith, J. R. (1952). *Min. Mag.* **29**, 953.
105. Coombs, D. S., Ellis, A. D., Fyfe, W. S. and Taylor, A. M. (1959). *Geochim. Cosmochim. Acts,* **17**, 53.
106. Breck, D. W. (1973). "Zeolite Molecular Sieves". Wiley, New York.
107. Gruner, E. (1934). *Z. anorg. Chem.* **219**, 181.
108. Gerrard, W. and Griffey, P. F. (1960). *J. Chem. Soc.* 3170.
109. Beekenkamp, P. and Stevels, J. M. (1963). *Phys. Chem. Glasses,* **4**, 229.
110. Leicester, J. (1948). *J. Soc. Chem. Ind.* **67**, 433.
111. Van Wazer, J. R. (1958). "Phosphorus and its Compounds", Vol. I, p. 553. Interscience, New York.
112. Kobeyashi, T. (1963). *Yogyo Kyokai Shi,* **71**, 201.
113. Greger, H. H. (1948). British patent 587 169.

6

Network Polymers of Mixed Connectivity Greater Than Four

Titanium, Zirconium, Tin and Cerium$^{(IV)}$ Orthophosphates and Orthoarsenates

The group IV elements titanium, zirconium and tin, together with cerium and thorium in their highest oxidation states form phosphates with the general formula $M(HPO_4)_2 \cdot xH_2O$, all of which are polymers with a layer structure. Each layer consists of an array of nearly coplanar, six-co-ordinate, group IV atoms linked through tetrahedral PO_4 groups lying alternately above and below the plane; each group IV atom is connected to six oxygen atoms belonging to six different phosphate groups, so that the polymer network has a connectivity that is partly four and partly six (Fig. 6.1). The layers are joined together by very

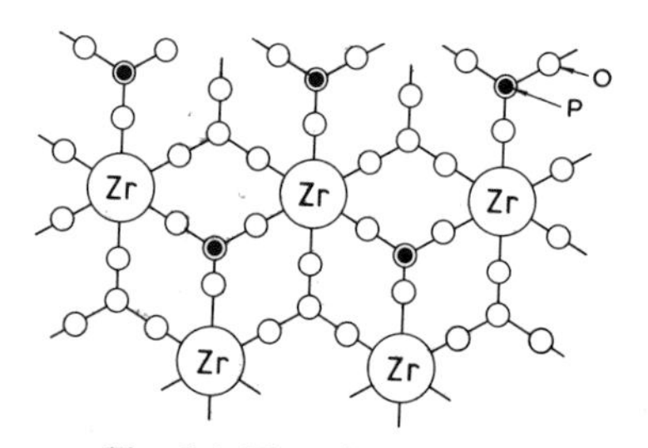

Fig. 6.1. Zirconium phosphate.

long hydrogen bonds between the non-bonding oxygens of the phosphate tetrahedra in adjacent layers. In one crystalline modification the phosphorus atoms in one layer lie opposite to the group IV atoms in the next layer; in other modifications the phosphorus atoms are opposite each other. In either case, zeolitic-like cavities are formed between the layers, connected to one another by openings of somewhat smaller size. There are usually water molecules in the centre of each cavity that are hydrogen-bonded to the phosphate groups, and the hydrogen ions that are associated with each PO_4 group are exchangeable for metallic cations. In addition to a number of crystalline modifications, all these polymers can be obtained in an amorphous form, in which the atomic proportions of group IV elements to phosphorus are freely variable; the amorphous materials are normally produced as aquagels which dry out to a porous glass. The first maximum in the X-ray powder pattern of the crystalline polymers is generally assumed to correspond to the interlayer distance, which varies with the size and charge of the interlayer cations, as well as with the nature of the group IV element. Interlayer distances for titanium, zirconium, tin, cerium and thorium phosphates, together with some of the isomorphous arsenates, are given in Table XVII.

Table XVIII.
Interlayer spacings in group IV phosphates and arsenates $M[H(P,As)O_4]_2 \cdot H_2O$

Group IV Element	Interlayer spacing in nm in monohydrates	
	Phosphates	Arsenates
Titanium	0·756	0·777
Zirconium	0·760	0·782
Tin	0·776	0·777
Cerium	1·095	0·910
Thorium	1·147[a]	0·705

[a] Trihydrate.

Titanium Phosphate

Titanium phosphate is formed by reaction of a titanium^{IV} salt, such as the chloride, with either orthophosphoric acid or an acidic solution of sodium dihydrogen phosphate. The precipitate formed initially is amorphous, and aqueous solutions of $TiCl_4$ and NaH_2PO_4 slowly form a gel which dries out to a granular, amorphous solid. Various techniques have been described for the preparation of amorphous powders[1,2] and granular forms[3,4] of titanium phosphate. These amorphous products have P:Ti ratios which can vary from about 0·6 to 2·0. Crystalline titanium phosphate with the composition $Ti(HPO_4)_2 \cdot H_2O$ is obtained by prolonged digestion of the amorphous form of the compound in orthophosphoric acid at 100°C or reflux, and the water molecule can be removed reversibly by heating to 120°C.

Titanium phosphate is resistant both to acids and to ionizing radiation, and for this reason has been used as an ion-exchange material for the separation of ^{137}Cs from strongly acidic solutions produced in the reprocessing of nuclear fuel elements,[5] and for the removal of radium from thorium and actinium at Curie-levels of activity.[6] Its theoretical ion-exchange capacity based on the formula $Ti(HPO_4)_2 \cdot H_2O$ is 7·76 mEq g^{-1} and values of around 7·5 mEq g^{-1} have been observed for the amorphous material[1] and 7·15 mEq g^{-1} for the crystalline form.[7] As an ion-exchanger it is easily regenerated and the amorphous variety has a high selectivity towards Cs^+. In the crystalline form, exchange of H^+ for Li^+ produces no change in the X-ray diffraction pattern, but the introduction of larger cations causes increases in the first diffraction maxima. Titanium phosphate also undergoes a unidimensional crystalline expansion when amines are introduced, either from the gas phase or solution.[8]

Glassy, amorphous titanium arsenate is prepared by mixing solutions of sodium arsenate with titanium tetrachloride.[9] This is also a cation-exchanger, but has a capacity of only about 1 mEq g^{-1}. Crystalline titanium arsenate of composition $Ti(HAsO_4)_2 \cdot 2H_2O$ has been prepared by dissolving freshly precipitated amorphous titanium arsenate in 8M arsenic acid and refluxing for 45h.[10] This product is much less stable towards acids than the phosphate and is extensively hydrolysed at pH > 5, so that its total ion-exchange capacity cannot be utilized.

Zirconium Phosphate

Zirconium phosphate has been more thoroughly investigated than any of the other compounds in this group. The earliest studies were mainly concerned with the amorphous forms, while more recent work has focused attention on the crystalline material. Zirconium phosphate aquagels are readily obtained by mixing zirconium salt solutions with phosphoric acid at ordinary temperature; products with Zr : P ratios from 0·5 to 2·1 have been obtained.[11] The aquagel can be granulated by freeze-drying the coarsely broken lumps of gel at −22°C. Crystalline zirconium phosphate can be prepared in two ways: firstly, by digestion or refluxing the amorphous material in phosphoric acid; and secondly by slow decomposition of a zirconium fluoro-complex in the presence of phosphoric acid.[12] In this procedure, a zirconium salt is dissolved in hydrofluoric acid and added to phosphoric acid. By slowly evaporating the hydrofluoric acid at room temperature or more quickly by warming at 80°C in a current of nitrogen, saturated with water vapour to avoid evaporating water, relatively large crystals of zirconium phosphate can be obtained. This yields a compound of formula $Zr(HPO_4)_2 \cdot H_2O$, the crystal structure of which has been established by Clearfield and Smith.[13] Its theoretical cation exchange capacity is 6·64 mEq^{-1} and this can be attained with all alkali cations at sufficiently high pH values. The anhydrous compound $Zr(HPO_4)_2$ has been obtained by drying the monohydrate over anhydrous calcium sulphate and also by heating at 130°C. A dihydrate was produced by air-drying the product of refluxing solutions of zirconyl chloride with sodium dihydrogen phosphate in hydrochloric acid. The exchangeable protons begin to be lost at about 200° and the cross-linking between the layers increases, so that the exchange capacity is reduced, but about half of the theoretical capacity is retained even up to 450°C, at which temperature a significant proportion of the pyrophosphate, ZrP_2O_7, is starting to be formed.[14]

The interlayer distances of several different salt forms of zirconium phosphate have been measured,[15] and it is found that the interlayer distance increases both with increasing radius of the cation and with increasing water content. The observed difference between the interlayer distances of two salts with the same degree of hydration is approximately equal to 1·1 times the difference between the diameters of the cations. Because the interlayer distances of several monohydrated salts, $Zr(MPO_4)_2 \cdot H_2O$, are the same as the corresponding anhydrous compounds, Alberti and Constantine[16] assumed that the water molecules in the monohydrate are located near the centre of the cavities between the layers, where there is sufficient space to accommodate them without expansion of the lattice. The monovalent cations must be situated around the water molecules and near to pairs of P—O⁻ groups in adjacent layers. On this basis, Alberti and Constantino concluded that the interlayer distance for the monohydrated forms of various salts of zirconium phosphate could be calculated from the radius of the cation according to the equation

$$d = 2{\cdot}8 + 3{\cdot}4 \cos\alpha + (1{\cdot}4 + r)^2 - 11{\cdot}56 \sin^2\alpha,$$

where r is the radius of the cation and

$$\alpha = 46{\cdot}9 - \text{arcos}\,(0{\cdot}95 + 0{\cdot}1\, r - 0{\cdot}036\, r^2).$$

The interlayer distances increase by 0·18 nm for every two additional water molecules that are present over and above the monohydrate.

Zirconium phosphate has been proposed as an ion-exchanger for purification of nuclear reactor coolants, decontamination of radioactive waste water, separation of ^{137}Cs from reprocessing solutions, and as a component of ion-exchange membranes for electrodialysis and in hydrogen-oxygen fuel cells.[17]

Tin^{IV} Phosphate

Amorphous tin phosphate aquagels are formed by mixing aqueous solutions of sodium dihydrogen phosphate and tin tetrachloride pentahydrate;[18] tin phosphate is also obtained by rapid addition of tin tetrachloride to aqueous orthophosphoric acid at room temperature when an amorphous precipitate is obtained. On heating with orthophosphoric acid, these amorphous products at first dissolve, and then reprecipitate in a crystalline form.[19] The crystalline product, after drying over phosphoric oxide, has the formula $Sn(HPO_4)_2 \cdot H_2O$, but on exposure to the atmosphere absorbs a further molecule of water. Heating at 100–300°C causes a reversible loss of water of hydration, but above 400°C the acid phosphate groups begin to condense with formation of the pyrophosphate.

Amorphous tin phosphates with P : Sn ratios from 1·25 to 1·5 are stable to heat and to ionizing radiation and their ion-exchange capacity increases with phosphorus content. The theoretical ion-exchange capacity of the crystalline polymer $Sn(HPO_4)_2 \cdot H_2O$ is 6·08 $mEq\,g^{-1}$ and this value is attained in sodium exchange, but in lithium exchange a higher capacity (7·9 $mEq\,g^{-1}$) has been reached, indicating the presence of more than two exchangeable protons per formula weight. This has been explained[20] by the following tautomerism in the solid:

$$\begin{array}{ccc} OPO_3H & & OPO_3H_2 \\ | \diagup & & | \\ Sn \leftarrow OH_2 & \rightleftharpoons & Sn - OH \\ | \diagdown & & | \diagdown \\ OPO_3H & & OPO_3H \end{array}$$

Stannic phosphate has been used to separate uranium from strontium, caesium and rare earths, and strontium and caesium from fission product solutions; and paper impregnated with stannic phosphate is useful for chromatographic separation of many inorganic cations.[17]

Stannic arsenate has been obtained in an amorphous form by mixing aqueous solutions of tin tetrachloride and either disodium hydrogen arsenate or arsenic acid; the product is stable to dilute acids and to relatively high temperatures. Depending on the arsenic content, the exchange capacity lies between 0·79 and 0·94 $mEq\,g^{-1}$. Crystalline tin arsenate is obtained in a similar way to tin phosphate, by refluxing the amorphous material in arsenic acid; the crystalline product $Sn(HAsO_4)_2 \cdot H_2O$ is easily hydrolysed by alkalies.[20]

Cerium^{IV} and Thorium Phosphates

Cerium and thorium phosphates are similar to the group IV B phosphates in chemistry and network structure, but differ markedly in their morphology. Both these compounds exist in a fibrous form as well as in microcrystalline and amorphous forms. As an example of a synthetic inorganic fibre, fibrous cerium phosphate is discussed in some detail in another section (Chapter 7) and will not be dealt with here. Amorphous cerium phosphates are formed when a cerium^{IV} salt is reacted in aqueous solution with cold phosphoric acid; the composition of these products varies, with P:Ce ratios from 1·03 to 1·95. When the precipitation is carried out in sulphuric acid, the P : Ce ratio decreases and the product contains from 2–6% of sulphate.[21] Amorphous cerium phosphate has an ion-exchange capacity of around 3 $mEq\,g^{-1}$, which is entirely lost on heating at 200°C. It also exhibits electron-exchange properties, oxidizing manganese^{II} ions to manganese^{III} and iodide to iodine.[22] Cerium phosphates have been used for separation of sodium from caesium, and of caesium from barium, in an ion-exchange column.[17] Cerium arsenate can be obtained by similar procedures to the phosphate, but has no fibre-forming tendency.

Thorium phosphate prepared by precipitation at low temperature has a variable composition, poor hydrolytic stability, and a low ion-exchange capacity. However, by precipitation at 100°C and subsequent digestion for 40 h, fibrous crystalline thorium phosphate, similar to fibrous cerium phosphate, is obtained.[23] Its ion-exchange capacity (3·7 $mEq\,g^{-1}$) is close to that expected for the empirical formula $Th(HPO_4)_2 \cdot 3H_2O$ and this material has been used to make experimental ion-exchange membranes.

Silicon Phosphate and Related Polymers

Silicon and germanium differ from the remaining group IV elements (other than carbon) in that the orthophosphates $Si(HPO_4)_2$ and $Ge(HPO_4)_2$ are much less stable, and it is the diphosphates (or pyrophosphates) SiP_2O_7 and GeP_2O_7 that are the usual products from the reaction of silica and germania or their halides with orthophosphoric acid. A compound with the composition $Si(HPO_4)_2$ has been obtained by heating freshly precipitated silica gel with phosphoric acid at 200°C[24] but this readily dehydrated to give SiP_2O_7 at higher temperatures.

Silicon phosphate has been known for a long time, but it has not always been recognized as a polymeric compound. As early as 1867, Skey[25] observed that molten sodium phosphate dissolves silica, and Laufer in 1878[26] recorded that molten sodium ammonium hydrogen phosphate attacked quartz at red heat. Hautefeuille and Margottet[27] found that freshly precipitated silica dissolved in phosphoric acid even at room temperature and that, on heating the solution, crystals having the composition $SiO_2 \cdot P_2O_5$ were deposited. The corrosion of quartz and vitreous silica by phosphoric acid was studied by Moore and Barton[28] and by Kosting and Heins.[29] More recently, the author[30] found that phosphoric acid began to attack the surface of glass at 300°C, at which temperature diphosphoric acid was being formed as the result of dehydration. Under certain conditions, the reaction results in the formation of barnacle-like outgrowths on the surface of the glass; these were found to consist of conical aggregates of needle-shaped crystals that could be isolated by dissolving away the unreacted glass in hydrofluoric acid. The insoluble residue was crystalline silicon phosphate, SiP_2O_7. The same polymer was obtained by Gerrard and

:ke[31] by polycondensation of triethyl phosphate with silicon tetrachloride:

$$2(EtO)_3PO + SiCl_4 \rightarrow SiP_2O_7 + 4\,EtCl + Et_2O.$$

Silicon phosphate is a crystalline network polymer in which every silicon atom is joined to six oxygen atoms and every phosphorus to four oxygen atoms, so that its connectivity is a mixture of four and six. The phosphorus atoms are joined in pairs through a shared oxygen atom to form $P_2O_7^{4-}$ units each of which shares its six remaining oxygen atoms with the SiO_6 octahedra (Fig. 6.2). The

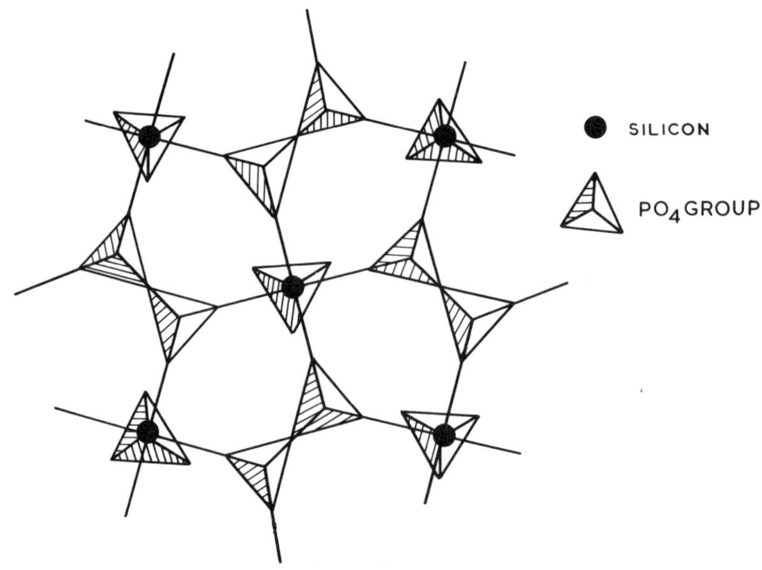

Fig. 6.2. Cross-section through the silicon phosphate network, showing the disposition of the pyrophosphate units.

density is 3·232 g cm^{-3}; its infra-red and Raman spectra have been recorded by Steger and Leukroth[32] and Hubin and Tarte.[33] Silicon phosphate is completely unaffected by strong acids, including hot concentrated sulphuric acid and 40% aqueous hydrofluoric acid,[30] and only slowly attacked by hot, concentrated caustic alkalies.

The group IV B elements titanium, zirconium and hafnium, as well as germanium and tin, also form crystalline network polymers of the general type MP_2O_7, which are isomorphous with silicon phosphate, and which have similar chemical inertness and thermal stability.

A number of related polymers containing only silicon, phosphorus and oxygen have been described, but their structures have not so far been elucidated. A compound with the empirical composition $2P_2O_5 \cdot 3SiO_2$ was obtained by dissolving freshly precipitated silica in orthophosphoric acid at 200°, then heating the solution rapidly to 900°C and keeping it for one hour at this temperature.(34) Another compound formulated as $Si(PO_3)_4 \cdot 2H_2O$ was obtained by heating dimethyl diethoxysilane, $Me_2Si(OEt)_2$, with phosphoric oxide or phosphoryl chloride and distilling out the volatile products up to 200°C.(35)

Zirconium pyrophosphate, polyphosphate and hypophosphate have been prepared by mixing solutions of zirconyl salts with sodium pyrophosphate in nitric acid, sodium polyphosphate in water, and sodium hypophosphate ($Na_2H_2P_2O_6$) in hydrochloric acid, respectively. These products are amorphous ion-exchange materials which are selective for certain cations.(17)

References

1. Piret, J., Henry, J., Balon, G. and Beaudet, C. (1965). *Bull. Soc. Chim. France,* 3590.
2. Dolique, R. and Quenoun, F. (1966). *Trav. Soc. Pharm. Montpellier,* **26**, 353.
3. Netherlands patent application 6 603 607 (1966).
4. Belgian patent 649 389 (1964).
5. Beaudet, C., Cremer, J. and Demaere, G. (1969). *Atompraxis,* **15**, 165.
6. Huys, D. and Baetslé, J. H. (1967). *Centre Energ. Nucl. (Brussels)* BLG-422.
7. Alberti, G., Cardini-Galli, P., Costantino, U. and Torracca, E. (1967). *J. Inorg. Nucl. Chem.* **29**, 571.
8. Michel, E. and Weiss, A. (1967). *Z. Naturforsch.* B, **22**, 1100.
9. Qureshi, M. and Nabi, S. A. (1970). *J. Inorg. Nucl. Chem.* **32**, 2059.
10. Alberti, G. and Torracca, E. (1968). *J. Inorg. Nucl. Chem.* **30**, 3075.
11. Amphlett, C. B. (1964). "Inorganic Ion Exchangers". Elsevier, Amsterdam.
12. Alberti, G. and Torracca, E. (1968). *J. Inorg. Nucl. Chem.* **30**, 317.
13. Clearfield, A. and Smith, G. D. (1969). *Inorg. Chem.* **8**, 431.
14. Mounier, F. and Winand, L. (1968). *Bull. Soc. Chim. France,* 1829.
15. Clearfield, A., Duax, W. L., Medina, A. S., Smith, G. D. and Thomas, R. (1969). *J. Phys. Chem.* **73**, 3424.
16. Alberti, G. and Costantino, U. (1974). *J. Chromatography,* **102**, 5.
17. Veselý, V. and Pekárek, V. (1972). *Talanta,* **19**, 219–262.
18. Inoue, Y. (1964). *J. Inorg. Nucl. Chem.* **26**, 2241.
19. Fuller, M. J. (1971). *J. Inorg. Nucl. Chem.* **33**, 559.
20. Costantino, U. and Gasperoni, A. (1970). *J. Chromatography,* **51**, 289.
21. König, K. H. and Meyn, E. (1967). *J. Inorg. Nucl. Chem.* **29**, 1153.
22. Larsen, E. M. and Cilley, W. A. (1968). *J. Inorg. Nucl. Chem.* **30**, 287.
23. Alberti, G. and Costantino, U. (1970). *J. Chromatography,* **50**, 482.
24. Makart, H. (1967). *Helv. Chim. Acta,* **50**, 399.
25. Skey, W. (1861). *Chem. News,* **16**, 187.
26. Laufer, E. (1878). *Ber.* **11**, 935.
27. Hautefeuille, P. and Margottet, J. (1883). *Comptes rend.* **96**, 1052.

28. Moore, B. and Barton, T. H. (1937). *J. Soc. Chem. Ind.* **56T**, 273.
29. Kosting, P. R. and Heins, C. (1931). *Ind. Eng. Chem.* **23**, 140.
30. Ray, N. H. (1970). *J. Non-crys. Solids,* **5**, 71.
31. Gerrard, W. and Jeacocke, G. J. (1959). *Chem. Ind.* 704.
32. Steger, E. and Leukroth, G. (1960). *Z. anorg. Chem.* **303**, 169.
33. Hubin, R. and Tarte, P. (1967). *Spectrochim. Acta.* **23**, 1815.
34. Jary, R. (1957). *Ann. chim.* [13] **2**, 58.
35. Kreshkov, A. P. and Karateev, D. A. (1959). *J. Appl. Chem. U.S.S.R.* **32**, 384.

7

Inorganic Polymer Technology

Synthetic Inorganic Fibres

A characteristic feature of polymeric materials is that they can usually be drawn into fibres, and inorganic polymers are no exception to this generalization; however, the processing of inorganic fibres is more difficult than organic fibres as a rule, because much higher temperatures are involved and the fibres produced are usually more brittle. Nevertheless a large number of inorganic fibres have been developed for different purposes. Their special properties of high elastic modulus, high softening temperature, resistance to fire and chemical inertness make them particularly suitable for thermal and acoustic insulation, plastics reinforcement, filtration media and fire-resistant fabrics.

Nearly all commercially important organic fibres are crystalline, but synthetic inorganic fibres can be either crystalline or amorphous. The latter category includes all the different kinds of glass fibre, mineral wools and fused silica. Although these materials form the bulk of commercially produced synthetic inorganic fibres, they will not be discussed here because the polymers from which they are derived have already been described in the preceding chapters. The former group of polycrystalline inorganic fibres includes the three polymeric oxides alumina, titania and zirconia, and a few unrelated compounds, such as potassium titanate and cerium phosphate which have unusual fibre-forming tendencies. All of these polymers are sufficiently different from the compounds that have been considered so far to warrant separate discussion.

Alumina, Titania and Zirconia Fibres

Polycrystalline alumina, titania and zirconia fibres have been produced by a variety of methods from aqueous solutions and suspensions of suitable precursors. Such processes make it possible to study a greater range of compositions in less demanding temperature conditions that can be achieved by melt-spinning, and to obtain different crystalline forms. Essentially two kinds of process have been developed for producing fibres from these refractory oxides. In the first, a fairly concentrated suspension of minute crystals of the oxide is formed into filaments, usually with the aid of temporary binder which may itself be an organic polymer. After evaporation of the solvent, the filaments are heated in air until the crystals sinter together to form an intact fibre, the organic binder burning off during this treatment. One difficulty that arises in a process of this kind is that of grain growth; at the sintering temperature recrystallization can occur resulting in a coarse texture and a weak fibre.

Polycrystalline alumina fibres of high purity have been produced from a "slip" of finely divided α-alumina particles averaging 0·3–2·5 μm in diameter, in water containing a high molecular weight polyethylene oxide as binder.[1] The dried filaments were heated to burn off the organic polymer and then to sinter the alumina particles into an intact fibre about 10 μm in diameter. Similar processes have been used to produce fibres from beryllium oxide,[2] quartz,[3] and zinc oxide.[4] As a result of recrystallization, zirconia fibres sintered at 1700°C were found to have an average grain size of 17 μm; similarly, fibres of γ-alumina heated at 1000°C are converted into α-alumina and the crystals grow to the full diameter of the fibre, giving it the appearance of a bamboo rod. It has been found that the addition of small amounts (~ 5%) of calcium oxide tends to stabilize the grain size and prevent this happening.[5]

This second kind of process that is used to produce refractory oxide fibres utilizes a soluble compound of the metal which can be thermally decomposed to form the oxide. A solution of suitable viscosity, which may also contain an organic additive as a temporary binder, is spun into filaments which are then heated to convert the precursor into the desired oxide. This can be regarded as a special case of polycondensation, in which the small cross-section and large surface area of the fibres enable the volatile by-products of the condensation reaction to be eliminated rapidly and completely without disrupting the filaments or producing voids. In one process of this type[6] spinnable solutions were obtained by solvolysis of silicon and aluminium compounds dissolved in aliphatic alcohols containing an acid; the addition of soluble inorganic polymers such as polyvinyl alcohol or polyethylene oxide was found to be beneficial in raising the viscosity of the solutions and improving their spinning characteristics.

In a process patented by I.C.I. for the manufacture of alumina fibres, a solution of aluminium oxychloride containing 2% by weight of polyvinyl alcohol is evaporated until the viscosity reaches 80 $\mathrm{N\,s\,m^{-2}}$, and extruded through spinnerets to form 3 μm diameter filaments which are then converted into polycrystalline alumina fibres by heating at 800°C. The resulting fibres are uniform in diameter, flexible and with a silky texture. Their tensile strength is

1000 $MN\,m^{-2}$, Young's modulus 100 $GN\,m^{-2}$, and they are stable up to 1400°C.[7] Zirconia fibres are produced in a similar way by evaporation of an aqueous solution of zirconyl chloride containing 2% by weight of partially hydrolysed polyvinyl acetate until the viscosity is at least 80 $N\,s\,m^{-2}$, and then extruding into 3–5 μm fibres. These are heated at 1000°C for 24 h to convert them into a polycrystalline zirconia fibre with a tensile strength of 700 $MN\,m^{-2}$, Young's modulus 100 $GN\,m^{-2}$, and stable to 1600°C. The zirconia fibres have exceptional chemical stability, being resistant to both strong acids and alkalies.

Alumina fibres produced under the trade-name "Saffil" are being used for high-temperature thermal insulation in the form of blankets, wool and paper, for reinforcement of composites, and for sterilizing filters. In aqueous media, they develop a positive zeta potential which makes them peculiarly suitable for the removal of negatively charged bacteria and yeast cells. The properties of "Saffil" fibres are given in Table XIX.

Table XIX
Properties of "Saffil" fibres

Properties	"Saffil" alumina	"Saffil" zirconia
Density, $g\,cm^{-3}$	2·5	5·4
Melting point °C	>2000	>2700
Tensile strength, $MN\,m^{-2}$	1000	700
Young's modulus, $GN\,m^{-2}$	100	100
Average diameter, μm	3	3–5
Surface area, $m^2\,g^{-1}$	10	10
Chemical resistance	Attacked by strong acids and alkalies	Resists strong acids and alkalies

Zirconia fibres have also been produced by spinning a fibre from a polyzirconoxane, $[—Zr(R_1R_2)—O—]_n$, where R_1, R_2 are organic groups, and burning off the carbon-containing groups in oxygen;[8] alumina and alumina-silica fibres can be made in a similar way from polyaluminoxanes and polysiloxanes,[9] and from a copolymer formed by hydrolysis of a mixture of triethyl aluminium and dimethyl dichlorosilane.[10]

Titania fibres have been produced by acidifying a solution of titanium tetraisopropoxide, $Ti(OCHMe_2)_4$, and concentrating the resulting solution to a viscosity of 50 $N\,s\,m^{-2}$. After ageing for 16 h, this solution was spun to give 12 μm diameter fibres which were then heated for 15 min. at 570°C. The fibres had a tensile strength of 0·65 $MN\,m^{-2}$.[11]

The spinning of an unsupported inorganic fibre precursor is difficult because the "green" fibres have little strength, and when an organic binder is added, a combustion step to remove it becomes essential. For this reason, it is not unreasonable to consider using an organic fibre as a support. Alumina and zirconia fibres have been produced in this way, using rayon fibre as the support which can be in the form of yarn or even woven cloth. The organic support fibre may with advantage be treated with an amine or with dimethylsulphoxide to cause it to swell and become more absorbent. It is then impregnated with a solution of an aluminium or zirconium salt in an organic solvent. After drying, the organic fibre is burned off and the inorganic residue is crystallized by controlled heat treatment.[12] Zirconia textiles are manufactured by this process by Union Carbide under the trade-name "Zircar".

The formation of fibres without any kind of spinning or extrusion process and without any fibrous support is a comparatively rare phenomenon, but there are at least four examples of this happening in the inorganic polymer field. They are the alkali metal hexatitanates $M_2Ti_6O_{13}$, cerium^{IV} phosphate, a crystalline form of aluminium silicate that is identical to a naturally occurring fibrous mineral, imogolite, and the hydrothermal crystallization of boehmite in fibrillar form.

Potassium hexatitanate, $K_2Ti_6O_{13}$, was first produced in fibrous form by hydrothermal reaction of titanium dioxide with caustic potash in the presence of water at 600–700°C and 3000 atm. pressure. It is also formed by recrystallization of prismatic crystals of $K_2Ti_6O_{13}$ from molten potassium chloride or a mixture of potassium chloride and fluoride; the reaction of titanium dioxide with potassium carbonate in this molten salt medium also results in fibres.[13] The simplest procedure involves calcination of a mixture of TiO_2 and K_2CO_3 at 600–1100°C in a continuous operation; the lumps of crude product are broken down and suspended in water, and the aqueous suspension is then macerated to produce a paper-like pulp. From this, felt, paper and other non-woven textile products can be obtained.[14] Potassium titanate fibres average 1 μm in diameter and may be up to several millimetres long; they are composed of bundles of aligned fibrils with a definite crystal structure. The melting point is 1370°C and the density is 3·2 g cm^{-3}. Because potassium titanate is opaque to infra-red radiation from 0·9–2·4 μm, the fibres have excellent high-temperature thermal insulation properties. Sodium, rubidium and caesium hexatitanates are isomorphous with the potassium salt and can all be obtained in fibrous forms by similar processes. When the alkali titanate fibres are melted and the melt is cooled, the substance crystallizes out in the form of coarse, prismatic needles; but these can be reconverted to the fibrous form by recrystallization from molten salts (usually KCl) at temperatures below 1000°C. The greater the difference between the temperature of crystallization and the melting point of the crystalline phase that is separating, the greater is the tendency for the preferred direction of growth to be emphasized; thus, longer fibres are to be expected at lower temperatures even though the rate of growth may be smaller.

Alkali hexatitanate fibres are stable to boiling water, but slowly dissolve in hot sulphuric and hydrochloric acids; at 1000°C they gradually lose alkali by volatilization of the oxide and are eventually converted into rutile.

Cerium^{IV} phosphate, like titanium, zirconium, and tin^{IV} phosphates (see Chapter 6) has a layer structure consisting of hexaco-ordinate Ce atoms linked together into puckered sheets by PO_4 tetrahedra, each of which carries an exchangeable proton; the first X-ray diffraction maximum is at 1·59 nm, which is approximately twice that observed for the corresponding zirconium and tin compounds. Its chemical composition is $Ce(HPO_4)_2 \cdot 3H_2O$, and although two molecules of water can be removed without significant change by desiccation over phosphoric oxide, the third molecule of water is lost only on heating to 180°C and its removal is accompanied by a marked deterioration in the fibrous structure and X-ray pattern. It has been suggested that the polymer molecules may be held together in the fibre by hydrogen bonds.[15]

Fibrous cerium phosphate is formed by the slow addition of a ceric salt (typically the sulphate dissolved in dilute sulphuric acid) to a hot (80–100°C) solution of orthophosphoric acid with constant stirring. The production of crystalline fibres is strongly dependent on temperature and reaction time[16] and the optimum conditions are reported to be 94°C and 4 h. At low temperatures, mainly amorphous material is precipitated, and at temperatures above 100°C, the product is crystalline but non-fibrous and particulate. Prolonged digestion in hot phosphoric acid also tends to give a non-fibrous product.

These fibres are obtained in bundles averaging about 50–100 μm long and 0·2–1·0 μm in diameter, consisting of aligned fibrils which are single crystals about 10 nm in diameter and up to 100 μm in length. The suspension is readily filtered on a gauze to form a paper-like material which has cation-exchange properties, and which has been used for chromatographic separations of inorganic cations and as an experimental ion-exchange membrane. Cerium phosphate has a very high selectivity for lead^{II} cations.[15] Thorium^{IV} phosphate can also be obtained in a fibrous form, but the corresponding arsenates cannot.

Imogolite is a naturally occurring crystalline aluminium silicate which is widely distributed, although in low concentrations, in volcanic deposits and is found as a minor component in soils derived from the younger igneous rocks; it was first recognized in Japan in 1962.[17] Its structure is believed to be tubular, the walls of the tubes being composed of a single sheet of gibbsite, $Al(OH)_3$, with orthosilicate groups replacing the hydroxyl groups on the inside surface of the tube. Its chemical composition is $(HO)_3Al_2O_3SiOH$. Fibrous imogolite has recently been synthesized by acidifying a dilute aqueous solution containing silicate and aluminium ions and maintaining this mixture close to 100°C for five days.[18] The product consisted of tubular fibres about 2·2 nm external diameter and 1·0 nm inside diameter which had the same infra-red absorption spectrum and electron diffraction pattern as the natural mineral.

Boehmite (AlO·OH) has also been made to crystallize in fibrillar form by hydrothermal treatment of hydrous alumina solutions. In a typical experiment,[19] aluminium powder was dissolved in M/1 aqueous aluminium chloride solution until the atomic ratio Al:Cl was 4:3, and the diluted solution was heated for 5 h at 160°C. A white precipitate formed, which consisted of compact bundles of boehmite fibrils, averaging 5 nm in diameter and up to 1 μm in length. Aqueous dispersions of this product gave clear, coherent films on drying, which survived heating to 950°C.

The value of inorganic fibres rests upon their thermal and oxidative stability (which implies non-flammability) and their intrinsically high modulus of elasticity. The additional fire hazard that results from the widespread use of synthetic organic fibres in everyday surroundings is too well known to need emphasis, and looking to the future it is not unrealistic to expect that wholly inorganic curtains, carpets and other soft furnishings may one day becone commonplace, if not actually mandatory. In these applications it is likely to be the spun filaments such as glass fibres that will be used to the greatest extent; but for reinforcement of composite materials where the principal requirement is high modulus, short staple fibre is as valuable as continuous filament provided its aspect ratio is large enough. Since the modulus of a fibre-reinforced composite is proportional to the product of. the modulus of the reinforcing fibre and its volume fraction in the composite, the relative value of different fibres as reinforcement is not determined by their absolute modulus, but by their specific modulus; that is their stiffness per unit volume. Values for some representative reinforcing fibres have already been given in Table I, and from these figures it is apparent that inorganic fibres in general suffer from the disadvantage of a comparatively high density. With traditional reinforcing fibres such as asbestos, glass fibre and steel filaments this disadvantage is to a large extent offset by a low price. The crystalline inorganic fibres all have a higher modulus than glass and in some cases (e.g. "Saffil") they can maintain their properties at very much higher temperatures, but because of their higher density and higher cost it is unlikely that any of them will displace glass fibre as a reinforcement for composites except where high-temperature performance is a critical factor.

Cement

Cement is probably the cheapest and the most widely used synthetic inorganic polymer. It consists mainly of anhydrous, crystalline calcium silicates, the major constituents (75% of Portland cement) being tricalcium silicate, Ca_3SiO_5, and β-dicalcium silicate, Ca_2SiO_4, in the ratio of approximately 2:1. Although there have been many extensive and detailed studies of the mechanism of setting of cement, it is still not completely understood. When anhydrous cement is mixed with water, the calcium silicates react to form hydrates and calcium hydroxide, a process which is inadequately described by equations such as

$$2\ Ca_3SiO_5 + 6\ H_2O \rightarrow Ca_3Si_2O_7 \cdot 3\ H_2O + 3\ Ca(OH)_2$$

$$2\ Ca_2SiO_4 + 4\ H_2O \rightarrow Ca_3Si_2O_7 \cdot 3\ H_2O + Ca(OH)_2.$$

Hardened Portland cement paste consists of roughly 70% calcium silicate hydrate, $Ca_3Si_2O_7 \cdot 3H_2O$, and 20% of crystalline calcium hydroxide, together with unhydrated calcium silicate particles, and smaller amounts of calcium aluminate, ferrites, calcium oxide, etc.

Since a typical cement paste consists of about 60% by volume of water and only 40% by volume of solids, the setting process must involve a considerable redistribution of the constituents of the original cement solids into the interstitial space occupied by water. Exactly how this happens is not yet known with certainty. The earliest attempt to explain the setting of cement was made by Le Chatelier[(20)] who suggested that the calcium silicate grains dissolved in water to give solutions that are supersaturated with respect to calcium silicate hydrate. Crystallization was thought to occur in the spaces between the grains to form an interlocking network. Later, Michaelis proposed that a gel was formed in the interstitial spaces, and since then opinions have alternated between the crystallite theory and the gel theory and various combinations of them both. The central problem still remains unsolved: what is the nature and mode of formation of the interstitial material in hardened cement? One recent view suggests that it is a polymeric material that is formed initially as fibrils that rapidly become enmeshed together, and if this is the case it could be that a re-examination of the structure of cement from the standpoint of polymer science may lead to significant improvements in the mechanical properties of cementitious materials such as concrete.

There is general agreement that the hydrated calcium silicate that is formed in the hydration of cement is of variable composition, and that it is poorly crystalline, if not actually amorphous. In the past there has been confusion arising from a failure to distinguish sufficiently clearly between the varieties of hydrated calcium silicates produced in the laboratory under carefully controlled conditions, and the substance that is actually formed in cement pastes. It has only recently become possible to use electron microscopy for observing the phenomena that occur during the hydration of cement, because of the necessity for maintaining a wet environment. Very high voltage microscopes now make it possible to incorporate cells with controlled environments within the microscope, and this facility has revealed a number of important facts that were hitherto unrecognized. Double and Hellawell[(21)] studied the hydration of cement grains in an electron microscope and showed that it is essentially a two-stage process in which a gelatinous layer is initially formed on the surface of the calcium silicate particles. After the initial setting time of about two hours, this gel layer begins to sprout fibrillar outgrowths which may actually be tubes, and which radiate from each grain into the interstitial space. This is very similar to the growth of "silicate trees" that are formed when a piece of metal or a crystal of a metal salt is placed in a solution of sodium silicate. The fibrils rapidly increase in length and number, and gradually become closely enmeshed. Eventually the lengthwise growth ceases and the fibrils begin to join up sideways, so that finally striated sheets of material are built up. During this process, calcium ions from the original calcium silicate particles are slowly dissolving, leading to the formation of calcium hydroxide in the aqueous phase. When the water becomes locally supersaturated with respect to calcium hydroxide, crystals of the hydroxide form and begin to grow, so that the interstitial material eventually becomes a dense mixture of calcium hydroxide crystals and hydrated calcium silicate gel. At this stage the morphology can no longer be clearly distinguished.

In cement that has set, the matrix material has a very high surface area, of the order of 200–300 $m^2 g^{-1}$, which clearly indicates a high porosity that is typical of a dehydrated gel.[22] Higgins and Bailey[23] studied the fracture of hydrated Portland cement paste and concluded that it was a notch-insensitive material, its strength being governed by its microstructure and hardly, if at all, affected by external flaws. This is what would be expected if the matrix consists of loosely connected plates, fibrils and tubes. Electron micrographs of cement pastes give evidence of a microstructure that is at least consistent with this description.[21]

Both calcium disilicate and calcium trisilicate have an orthosilicate structure, consisting of a framework with channels large enough to admit of penetration by water molecules. Hydrolysis produces calcium ions and hydroxide ions which pass into the solution, leaving orthosilicic acid on the surface of the calcium silicate crystal:

$$Ca_2SiO_4 + 4\ H_2O \rightarrow 2\ Ca^{++} + 4\ OH^- + Si(OH)_4.$$

This suggests that the initial product of hydration of the surface of the cement grains is probably orthosilicic acid, which will gradually polymerize to form a gel. Tamas[24] has confirmed that the anhydrous calcium silicates present in cement are orthosilicates by the procedure of trimethylsilylation of the anions and identification of the products by gas chromatography, and he has also shown that in the early stages of hydration, the products are mainly dimeric silicic acid, with a little trimer and tetramer. The monomeric anion SiO_4^{4-} is only stable above pH 13·6, while between pH 10·9 amd 13·6, the principal component is the dimer $Si_2O_7^{6-}$. Since the equilibrium pH of the aqueous phase of hydrating cement is 12·7, the orthosilicic acid released by hydrolysis of the anhydrous calcium silicates will polymerize rapidly, initially to the dimer; this is followed by coagulation of the disilicate by calcium hydroxide, forming a membrane around the hydrating particle. As hydration continues at the surface of the particle the concentration of calcium ions in the solution inside the membrane rises, causing water to diffuse into the envelope under osmotic pressure, and eventually the membrane bursts open in several places releasing fresh disilicate anion, which again coagulates, building a hollow tubular structure similar to a "silicate garden", but inside out in relation to the usual schoolboy's experiment. When there is enough space between the grains, as in a very dilute cement paste, this tubular morphology can be seen by electron microscopy; but in a paste of the usual concentration for a practical cement, the morphology is distorted by mutual interference and lateral combination of the protuberances from adjacent grains.

Following the initial set, there is a slow increase in strength over a longer period of time, and Lentz[25] has shown that this is accompanied by a slow polymerization of the disilicate and trisilicate anions to higher polymers, increasing both in quantity and in molecular weight with time. Thus the matrix which eventually binds the residual cement particles together is a hydrated polysilicate glass. This provides an explanation of the effect of alumina in cement. At the high pH of a cement paste, any aluminium will be present in the form of aluminate anions, and these can substitute as AlO_4 tetrahedra for some of the SiO_4 units in the polysilicate network. This substitution produces an acid site in the network which will accelerate the polycondensation reaction, causing a faster built-up of the covalent network and consequently a more rapid increase in mechanical strength for similar setting times.

It remains to explain why cement, with an apparently "continuous" polysilicate matrix, has such a low tensile strength compared to glass. The polysiloxane network in a silicate glass in continuous and, with special surface treatment, its tensile strength can reach 17 GN m^{-2}. Normally, surface flaws limit its strength to around 60–70 MN m^{-2}, and this is the value that might be expected for a cement paste with an intact and continuous silicate matrix. In practice, however, the tensile strength of cement rarely exceeds 15 MNm^{-2}, and not one of a very large number of different additives that have been proposed to improve the strength of cement, including substances intended to entrain air, repel moisture, accelerate or retard setting, etc.,[26] will improve the ultimate strength beyond 20 MN m^{-2}. The only exception to this is the incorporation of water-soluble polymers that modify the rheology of the paste, enabling it to be extruded in sheet form which, after carefully controlled hydration and drying, can attain a tensile strength of 30 MN m^{-2}. The essential feature of this process which has been developed by I.C.I.[27] is a considerable reduction in the initial water content of the paste, and hence of the final porosity of the set cement. It is practically certain that the relatively high volume fraction of voids in ordinary cement is responsible for its low tensile strength compared to glass, and it is reasonable to expect that further reduction in the porosity might ultimately lead to cementitious materials with the same order of tensile strength as ordinary glass.

References

1. German patent 2334704 to Aluminium Company of America.
2. British patent 1 001 003 to National Beryllia Corporation.
3. U.S. patent 3 177 057 to Engelhard Industries Inc.
4. U.S. patent 3 703 413 to McDonnel Douglas Corporation.
5. Economy, J. (1966). "Encyclopaedia of Chemical Technology", Vol. 11, p. 651. Interscience, New York.
6. British patent 1 312 716 to Bayer A-G.
7. British patents 1 360 197; 1 360 198; 1 360 199 and 1 360 200 to Imperial Chemical Industries.
8. Japanese patent 49/134928 to Sumitomo Chemical Industries.
9. German patent 2 408 122 to Sumitomo Chemical Industries.
10. Japanese patent 50/18726 to Sumito Chemical Industries.
11. German patent 2 418 027 to Minnesota Mining and Manufacturing Co.
12. Hambling, B. H. *et al.* (1973). Sintering and Related Phenomena, *in* "Material Science Research" (Ed. Kuczynski, G. C.) Vol. 6, p. 425. Plenum Press, London.
13. Berry, K. L., Aftandilian, V. D., Gilbert, W. W., Meibohm, E. P. H. and Young, H. S. (1960). *J. Inorg. Nucl. Chem.* **14**, 231.
14. Ovechkin, E. K. (1968). *Izv. Akad. Nauk S.S.S.R., Neorg. Mat.* **4**, 1141.
15. Alberti, G. and Costantino, U. (1974). *J. Chromatography*, **102**, 5.
16. Alberti, G., Costantino, U., di Gregorio, F., Galli, P. and Torracca, E. (1968). *J. Inorg. Nucl. Chem.* **30**, 295.
17. Wada, K., Yoshinaga, N., Yotsumoto, H., Ibe, K., and Aida, S. (1970) *Clay Minerals,* **8**, 487.
18. Farmer, V. C., Fraser, A. R. and Tait, J. M. (1977). *Chem. Comm.* (13), 462.
19. Bugosh, J. (1961). *J. Phys. Chem.* **65**, 1789.
20. Taylor, H. F. W. (1964). "The Chemistry of Cements", Vol. 1. Academic Press, London.
21. Double, D. D. and Hellawell, A. (1976). *Nature,* **261**, 486.
22. Birchall, J. D., Howard, A. J. and Bailey, J. E. (1978). *Proc. Roy. Soc.* A **360**, 445-453.
23. Higgins, D. D. and Bailey, J. E. (1976) *J. Materials Sci.* **11**, 1995.
24. Tamas, F. D. (1976). "Hydraulic Cement Pastes: Their Structure and Properties". Proceedings of a Conference at the University of Sheffield, 8–9 April 1976.
25. Lentz, C. W. (1966). Special Report 90, p. 269. Highway Research Board, Washington D.C.
26. "Encyclopaedia of Chemical Technology" (1949). Vol. 3, p. 489. Interscience, New York.
27. British patent 1 405 090 (1975) to Imperial Chemical Industries.

8

The Future for Inorganic Polymers

The Outlook for Inorganic Materials

During the next twenty years or so up to the end of the present century we are likely to see a dramatic, though gradual, movement away from many of the organic polymers that we have grown accustomed to using, towards a new generation of materials that will be largely or even entirely inorganic. Some of these will be direct substitutes for the organic materials they replace, for example inorganic paper is already a practical possibility; but many more will be materials with entirely different combinations of properties which will necessitate new methods of processsing and a new technology for their effective utilization. One example in this category is glass fibre-reinforced cement.

There are two main reasons for this change. The first, which is likely to be the more important in the short term, is the combined result of a number of environmental factors such as health and safety and a growing desire to try to preserve as much as possible of what is left of the earth's natural scenery and climate. The inflammability of organic plastics has already stimulated extensive legislation in many countries to control their use in furniture and buildings, and the ensuing regulations are continually being made more and more stringent. The recent history of research for fire-retardant additives to control the flammability of plastics clearly shows that there is never likely to be a complete solution of this kind to the problem, and moreover all the most effective fire-retardants are either toxic themselves, for example tris(2,3-dibromopropyl)phosphate is now known to be a potent carcinogen, or give rise to toxic products such as antimony halides during combustion. Even plastics which are so difficult to ignite as to be almost non-inflammable are by no means free from hazard; in a fire involving large quantitites of polyvinylchloride, for example, the corrosion of steelwork by the hydrogen chloride evolved in the decomposition of the plastic has been sufficiently serious on some occasions to render what would otherwise have been a slightly damaged building totally unsafe.

Another environmental factor is the relatively high resistance of plastic waste to the natural processes of decomposition and dispersal. It is not just the unsightliness of litter that is largely made up of plastic cartons, wrapping films and bags that is important, but the length of time that elapses before such litter decays and disappears, and the harm it can do to animals and vegetation before it is removed. Broken glass is equally unsightly and in the short term can be just as hazardous, causing wounds and even starting fires; but the sharp edges of broken glass quickly become rounded by the combined action of the wind (producing abrasion) and the rain (causing hydrolysis), and, once it has broken down into small enough fragments, glass introduces no foreign constituents into the soil. It is for these and similar reasons that some polymer scientists have recently begun to study the synthesis and properties of bio-degradable polymers, and polymers with inbuilt decay mechanisms that can be triggered by sunlight.

The second and more important reason for the gradual but inevitable disappearance of the commoner organic plastics is the long-term exhaustion of the natural gas and oil deposits from which most of the relevant chemical feedstocks are derived. At the end of 1976 the world's proven oil reserves were $8{\cdot}83 \times 10^{10}$ tonnes (a drop of $9{\cdot}4 \times 10^{9}$ tonnes from 1974) and the known sources of natural gas amounted to $6{\cdot}58 \times 10^{13}\,m^3$.[1] The current (1976) annual consumption of oil in the whole world is about $2{\cdot}9 \times 10^{9}$ tonnes, giving a ratio of proven reserves to annual consumption of 30 years; the corresponding figure for 1974 was 34 years.

The rate at which new oil deposits are being discovered was declining up to 1969, and although since then it has remained substantially constant, it is barely keeping pace with the annual increase in consumption. As a preliminary estimate therefore, the earth's oil reserves may be expected to last another 30 years. A more reliable basis for predicting the rate of exhaustion of the earth's oil resources has been discussed by Hubbert.[2] The history of oil discovery and production in the United States over the period from 1900 to 1970 shows that the rate of change in production follows a similar pattern to the rate of discovery of new deposits with a time-lag of about 12 years. For the past 40 years the curve of cumulative total production has faithfully followed that of total proven reserves with a time lag that was rarely outside the range 10–12 years. This means that the history of world oil discovery can give a 12-year preview of the changes to be expected in world oil production. From observations on small areas from which oil deposits are becoming or have already been exhausted it is known that the peak rates of discovery and production both occur at about the mid-points of their total time-span. Assuming this to be also true of the whole world, it is possible to form an estimate of the total quantity of oil that can be extracted from the earth. Various estimates by this and by other methods[3, 4] give results that fall within the range $2 \cdot 5 \times 10^{11}$ to $3 \cdot 6 \times 10^{11}$ tonnes. On this basis, half the world's total supply of oil will have been used up by the end of the 1980s and the rate of production will begin to fall after that time. This will result in a serious shortage of hydrocarbon feedstocks for the chemical industry and a dramatic increase in the price of organic plastics. After the end of the twentieth century, synthetic organic polymers will have to be based on raw materials derived either from coal, or from renewable resources such as wood and cotton. Long before this, however, organic plastics will have become so costly that alternative materials will have had to be developed.

Extensions of Present Technology

The first step towards the eventual replacement of plastics by inorganic materials is likely to be the development of composite materials containing very much higher proportions of inorganic fillers than can at present be incorporated into thermoplastics. The proportion of a rigid, solid filler that can be mixed with an organic polymer is limited by a number of factors, of which the most important are the viscosity of the melt and the brittleness of the solid composite. With existing technology it is not practicable to mould articles of a filled polymer that contains more than about 30% by volume of rigid particles, and higher volume fractions of most fillers make the resulting plastic too brittle for many applications. Were it possible to increase the inorganic filler content of an organic plastic beyond 60% by volume without incurring the penalties of brittleness and processing difficulty, the fire-hazard associated with the resulting material could be considerably reduced, and its cost in terms of the oil needed for its production would be significantly less. If the inorganic filler is a substance that undergoes an endothermic change—either a simple phase change, or better still, an endothermic dehydration—at a temperature below the ignition temperature of the plastic, such a composite can be made to be virtually non-inflammable. For use in thermoplastics, such a filler would have to be stable at the processing temperatures involved, and because of the high loadings of filler that will be needed, the processing temperature is likely to be much higher than is used at present, so the lower limit for decomposition of the filler is probably in the region of 300°C. The temperature of ignition of thermoplastic materials can vary greatly, not only with composition, but also with conditions, but is rarely much above 500°C. Consequently the kind of filler that might be considered in this context is one that has a large endothermic change in the range 300–500°C. One example of such a compound is hydrated alumina, which begins to lose combined water at 330°C, and absorbs heat to the extent of 3 kJ g^{-1} in the process of dehydration. A burning zone will only propagate continuously through the material so long as the heat required for decomposition of the organic polymer into volatile products can be supplied by its heat of combustion. The heat of combustion of polyethylene, for example, is 40 kJ g^{-1} and the heat required for its decomposition into volatile products is approximately 2 kJ g^{-1}, leaving a surplus of 38 kJ g$-^{1}$. Consequently if there were 13 g of alumina trihydrate for every gram of polyethylene in a composite, it could not sustain combustion. Since the density of gibbsite is 2·42 g cm^{-3}, this would require a volume fraction of filler of about 0·8. Such a packing of filler particles is not only far beyond the scope of present-day filler technology, with uniformly sized spherical particles it is physically impossible, because the theoretical packing fraction for uniform spheres in hexagonal close packing is 0·74, while in random packing, the maximum attainable packing fraction is only 0·63.

One way of increasing the proportion of any particulate filler than can be incorporated into a composite is to select a particle size distribution that permits much higher packing densities to be reached. A mixture of spheres with diameter ratios 1:7:38:316 in the proportions 6:10:23:61% by volume has been shown to attain a packing fraction of 0·95 and still to be sufficiently free-flowing to be poured from one container to another.[(5)] At a loading of 50%, a filler with this distribution of particle sizes produced a smaller increase in the viscosity of a polyester resin than a 37·5% loading of uniformly sized particles.[(6)] Thus with proper design, it should be possible to manufacture composites with a very high inorganic content that will be practically non-burning.

The problems associated with processing a very highly filled thermoplastic can sometimes be avoided by mixing the filler particles with liquid monomer or with a solution of polymer in monomer, and finally completing the polymerization of the organic part of the composite after shaping. This represents a processing sequence that is analogous to the production of ceramic ware and pottery, except that the "firing" temperatures will be very much lower. Another way of overcoming the high viscosity of a heavily filled molten polymer is to use a thermoplastic filler. Inorganic phosphate and borophosphate glasses have been prepared with softening temperatures as low as 200°C (Chapters 3 and 4), but at room temperature these materials can have elastic moduli that are only slightly lower than the silicate glasses that are used as reinforcement in plastics. A glassy polyphosphate of this kind with a glass transition temperature of 184°C has been incorporated as a reinforcing filler into polypropylene at loadings up to 60% by volume without greatly affecting the processability of the composite. In fact, the process of extrusion converted the originally particulate filler into fibres;[(7)] in one example a mixture of polypropylene granules and glass granules in the volumetric proportions of 3:2 was extruded through a die at 240°C to give a continuous rod 4 mm in diameter. The tensile modulus of this composite was seven times greater than that of unfilled polypropylene, and by extracting the hydrocarbon polymer from the composite with hot xylene, the glass was found to be in the form of 3–6 μm diameter fibres with an aspect ratio greater than 100, aligned in the direction of extrusion. Extrusion and rolling the same mixture into sheet form produced a composite with a laminar structure in which the glass was present as thin flakes.

A third way of avoiding the processing problems associated with thermoplastic composites containing very high loadings of inorganic particles is to reduce the size of the particles to molecular dimensions. With ordinary fillers there is an optimum size range for purely mechanical reinforcement, the lower end of which is about 18 nm, as exemplified by carbon black and some varieties of colloidal silica. If, however, reinforcement is not the primary objective, and the maximum possible inorganic content is required, then reducing the size of the filler particles almost to the same dimensions as the organic polymer molecules will result in an entirely new type of composite which is best regarded as a polymer blend. Wholly organic polymer blends can have valuable properties not achievable in any homopolymer or random copolymer, and materials such as ABS (acrylonitrile–butadiene–styrene polymers) have been extremely successful, but most polymer technologists have up to now ignored the possibility of developing polymer blends in which one of the components is inorganic. Inorganic particles in the required size range for this type of approach cannot, of course, be produced by any process of comminution of larger particles. Instead they will have to be synthesized, preferably, though not necessarily, in the presence of the organic component. One way in which this might be achieved is by preparation of a stable silica sol (for example, by acid hydrolysis of ethyl silicate) in a polymerizable organic monomer[8] and then polymerizing the monomer.

The Need for New Technology

Entirely new methods of processing will have to be developed for polymeric materials that are wholly inorganic. Apart from glass, the oldest of all the thermoplastics, and composite materials like ceramics and cement, inorganic polymers are not processable by the familiar methods of current technology. One kind of processing operation that seems likely to become important in the future might be termed "reconstructive processing". Whereas melt processing of a cross-linked polymer such as a silicate involves breaking down its network structure sufficiently to obtain viscous deformation, "reconstructive" processing consists in essence of carefully dissecting a naturally occurring mineral into minute, but geometrically similar fragments. Such fragments will usually be of colloidal dimensions, so that they are infinitesimally small in comparison with the article to be fabricated, but still one or two orders of magnitude larger than a polymer molecule. These fragments are then re-assembled into the shape required. An example of reconstructive processing is the production of sheet and film from clay minerals such as vermiculite. Walker and Garrett[9] showed that exchange of the interlayer magnesium cations in vermiculite for lithium[10] or *n*-butylammonium[11] cations causes the crystals to imbibe water and swell unidimensionally about 4–5 times in volume. The swollen mineral can then be dispersed in water by shearing forces, for example in a colloidal mill, to give an aqueous suspension of platelets that are only one or two silicate layers thick and tens of micrometres across. The peculiar morphology of such suspensions is reflected in their marked streaming birefingence, "ripple" structures, and interference colours. By evaporating layers of these suspensions on a suitable support, coherent films are obtained which can be built up in thickness by deposition of successive layers, and which can be stripped off the support to give flexible sheets. This film-forming ability of vermiculite dispersions is due to the morphology and structure of the individual platelets. When these are deposited by slow evaporation of the water, they tend to lie flat on their basal planes and overlap one another, so that strong electrostatic forces bind successive layers together to form a coherent sheet. Because the separate platelets are no more than about 1nm thick, any voids in the sheet are smaller (in a direction normal to the sheet) than the wavelengths of visible light, so that a properly reconstructed sheet of vermiculite can be quite transparent. In their freshly formed state the sheets are easily disintegrated by water, but replacement of the interlayer cations by magnesium so as to restore the original vermiculite composition renders the material stable to water and organic solvents. Such sheets, being wholly inorganic, are non-burning and retain their flexibility up to about 400°C.

Wholly inorganic composite materials are likely to become more and more important in the future. Another new development in processing and fabrication that is particularly relevant to inorganic composites involves the formation of crystals with chosen morphology within a composite. This is well illustrated by the mica glass-ceramics. While glass-bonded mica sheets and organic polymers loaded with mica have been known for a long time, these materials tend to be coarse grained, mechanically weak, and require a comparatively difficult forming operation to fabricate complex shapes. Recently a group of glass-ceramics has been developed that after crystallization contain a high concentration of mica crystals.[12, 13, 14] They are strong, resistant to both thermal and mechanical shock, and can be sawn and machined in a way that is impossible with homogeneous glass. One typical mica-forming glass ceramic[15] is melted from an aluminoborosilicate composition containing magnesia, potash, and fluoride ion. These glasses become opalescent on cooling due to liquid–liquid separation, and on re-heating at 700°C, the disperse phase undergoes a complex crystallization sequence that eventually results in the formation of fluorophlogopite mica. Like all the micas, crystals of this material have a low cleavage energy but they are very strong parallel to the sheet direction. Consequently fractures in such a composite propagate along the cleavage planes of the crystals. The high volume fraction and random orientation of the crystalline phase causes any crack that is initiated from the surface to follow an extremely tortuous path through the material, with frequent branching so that its energy is quickly absorbed; as a result the material has a fracture surface energy between 10 and 40 $J\,m^{-2}$, and it can be drilled and machined to precise dimensions with conventional metal-working tools.

In another type of glass-ceramic with a crystalline phase of β-wollastonite, the morphology of the disperse phase can be altered from fibrous or needle-like to granular, according to the temperature of heat treatment.[16] The composite with a fibrous disperse phase has better resistance to impact and a higher flexural strength.

The Need for Basic Research

Although motor-cars were running on rubber tyres long before anyone had begun to study the fundamental basis of rubber elasticity, basic research into the structure and properties of natural rubber has led to a better understanding of its behaviour, and has resulted in very considerable advances in the quality of the tyres that are now available. In the same vein, the fact that the branch of polymer science that is concerned with highly cross-linked inorganic systems has, until recently, been entirely neglected has not prevented us from constructing satisfactory buildings from reinforced concrete, or from making very considerable advances in glass technology. Nevertheless, it is quite certain that a better understanding of the principles underlying structure-property relationships in inorganic materials will not only result in a more effective utilization of the materials we have available at the present time, but will also lead to the discovery of new materials with improved properties. Because research of this nature necessarily takes a long time to be fruitful, we cannot afford to wait until the crisis point in the supply of organic materials arrives; this research must be intensified and extended now if the results are to be available in time.

References

1. "BP Statistical Review of World Oil Industry" (1976). British Petroleum Co., London.
2. Hubbert, M. King (1969) *In* "Resources and Man, A study and Recommendations by the Committee on Resources and Man of the Division of Earth Sciences", Chap. 8. W. H. Freeman, San Francisco.
3. Weeks, L. G. (1958). *Amer. Assoc. Petrol, Geol. Bull.* **42**, 431.
4. Hendricks, T. A. (1965). *US Geol. Survey Circular,* 522.
5. McGeary, R. K. (1961). *J. Amer. Ceram. Soc.* **44**, 513.
6. Ritter, J. (1971). *Appl. Polymer Symposium,* no. 15, 239–261.

7. Ray, N. H. and Sherliker, F. R. (1971). British patent 1 356 919 to Imperial Chemical Industries.
8. Deutsche Gold u. Silber Scheideanstalt, French patent 1 529 058 (1968).
9. Walker, G. F. and Garrett, W. G. (1967). *Science,* **156**, 385.
10. Walker, G. F. and Milne, A. A. (1950). *Trans. 4th Int. Cong. Soil Sci., Amsterdam,* **2,** 62.
11. Walker, G. F. (1960). *Nature,* **187**, 312.
12. Beall, G. H. (1972), *In* "Advances in Nucleation and Crystallization in Glasses" (Ed., Hench, L. L. and others), p. 251, American Ceramic Society, Ohio.
13. Chyung, C. K., Beall, G. H. and Grossman, D. G. (1972) *In* "Electron Microscopy and Structure of Materials" (Ed., Thomas, G. and others), p. 1167, University California Press, Berkeley.
14. Chyung, C. K. (1974). *In* "Fracture Mechanics of Ceramics" (Ed., Bradt, R. C.), p. 495. Plenum Press, New York.
15. Chyung, C. K., Beall, G. H. and Grossman, D. G. (1974). *In* "Xth International Congress on Glass", part II, pp. 14–33. Ceramic Society of Japan, Kyoto.
16. Kawamura, S., Yamanaka, T., Toya, F., Nakamura, S. and Ninomiya, M. (1974). *In* "Xth International Congress on Glass", part II, pp. 14–68. Ceramic Society of Japan, Kyoto.

Author Index

Numbers in parentheses are the reference numbers on text pages; numbers set in *italic* type are those pages at the end of chapters where references are listed in full.

L

M

N

O

P

Q

R

S

T

Subject Index

A

B

C

V

W

X

Y

Z